Compact Textbooks in Mathematics

This textbook series presents concise introductions to current topics in mathematics and mainly addresses advanced undergraduates and master students. The concept is to offer small books covering subject matter equivalent to 2- or 3-hour lectures or seminars which are also suitable for self-study. The books provide students and teachers with new perspectives and novel approaches. They may feature examples and exercises to illustrate key concepts and applications of the theoretical contents. The series also includes textbooks specifically speaking to the needs of students from other disciplines such as physics, computer science, engineering, life sciences, finance.

- **compact:** small books presenting the relevant knowledge
- **learning made easy:** examples and exercises illustrate the application of the contents
- **useful for lecturers:** each title can serve as basis and guideline for a semester course/lecture/seminar of 2-3 hours per week.

Franz Chouly

Finite Element Approximation of Boundary Value Problems

 Birkhäuser

Franz Chouly
Center of Mathematics
University of the Republic
Montevideo, Uruguay

ISSN 2296-4568   ISSN 2296-455X   (electronic)
Compact Textbooks in Mathematics
ISBN 978-3-031-72529-6   ISBN 978-3-031-72530-2   (eBook)
https://doi.org/10.1007/978-3-031-72530-2

This book is published under the imprint Birkhäuser, www.birkhauser-science.com by the registered company Springer Nature Switzerland AG
The registered company address is: Gewerbestrasse 11, 6330 Cham, Switzerland

If disposing of this product, please recycle the paper.

To my daughter

Foreword

Simulations have overtaken today's world, and whilst physicists and philosophers are busy seeking a "model of everything", engineers work hand-in-hand with mathematicians, to make those models reliable and useful to better understand our world and the inventions we create.

In this book, Professor Franz Chouly, whom I am fortunate to count as a close friend, focuses his attention to one particular numerical method known as the "finite element method" (FEM). The roots of FEM can be traced back to the early twentieth century. In the 1940s and 1950s, engineers working on structural analysis problems, particularly in the aerospace industry, began to develop methods to discretize continuous structures into simpler, discrete elements. This was a practical necessity due to the complexity of solving real-world engineering problems.

Finite element methods have continuously evolved through the concerted effort of engineers, on the one hand, and mathematicians, on the other. The former have been devising novel methods to tackle increasingly challenging problems, while the latter have been creating a mathematical edifice for those methods to stand fast and scale to real-life applications. Sometimes, these mathematical surgeries led to significant modifications of the original idea, at other times, mathematics and engineering have collaborated within a self-strengthening formal framework.

This continuous cross-fertilization of both fields started in the early pioneering days, with Walter Ritz and Richard Courant (1943), Alexander Hrennikoff and John Argyris (1950s), and Ray W. Clough (1960). Ivo Babuška worked on error estimates and convergence and Gilbert Strang facilitated building a continuum between mathematical bases and engineering practice. Later, Philippe Ciarlet and Pierre Arnaud Raviart from the French school of finite elements continued the theoretical efforts, bringing rigour in increasingly complex problem settings. By blending mathematical rigor with practical insights, Prof. Chouly does today for modern topics in finite element methods what Prof. Gilbert Strang and Prof. Zienkiewicz did in the 1970s for this nascent field.

He pays particular attention to boundary conditions, which are often superficially treated, although a topic of central importance, in particular when those are uncertain. This is particularly true of modern applications of the methods, such as biomechanics or for digital twins of engineering systems. Each chapter begins with a clear, informal outline designed to inspire further exploration. This approach lowers the barrier to entry, making complex concepts accessible without sacrificing depth. F. Chouly's method of demystifying the subject matter while maintaining its elegance is a testament to his dedication to both teaching and advancing the field.

For students, researchers, and professionals alike, this book offers a comprehensive guide to understanding and implementing the Finite Element Method. It bridges the gap between theory and practice, providing a robust framework for addressing some of the most challenging problems in numerical analysis. What sets this book apart is its dual focus: while rooted deeply in the mathematical foundations of FEM, it never loses sight of real-world applications. The discussions on well-posedness and error estimates are not merely theoretical exercises but are illustrated with practical implications. This makes the book invaluable not only for those new to the field but also for seasoned practitioners looking to deepen their understanding.

Prof. Chouly shows, in this book, his usual visionary abilities, by realizing the importance of a continuous blend between mathematical analysis and its applications to engineering and natural systems. As increasingly complex systems are being addressed computationally, the need to understand the various sources of error will become more important than ever. As data and model-based approaches blend together, this book, focusing rigorously on the quantification of discretization errors will become of paramount importance. As such, I am convinced that this didactic book is likely to become a reference for generations to come.

Esch sur Alzette, Luxembourg   Stéphane Bordas
June 2024

Preface

it is so difficult at present to decide whether functional analysis belongs to pure or to so-called applied mathematics (Lothar Collatz, preface to "Functional analysis and numerical mathematics", 1966)

The Finite Element Method, as a method to approximate the solution of some partial differential equations, has a long history behind, that goes back, at least, to the 1950s. Since then, many reference books have been devoted to this method, as well as many introductive books, with emphasis either on the mathematical foundations or on practical implementation issues for real-life problems, and sometimes both.

This present book is another introduction to the mathematics behind the Finite Element Method, and, in this aspect, it does not differ from many other textbooks. It belongs to a tradition that considers numerical analysis as a special branch of classical analysis, with foundations from the functional analysis of the twentieth century. This tradition can go back to the works of Lothar Collatz and Leonid Kantorovich.[1] This spirit has been present in many works in the finite element community, since the pioneering works of Ivo Babuška, Franco Brezzi, Claess Johnson, John Tinsley Oden, and many others, as well as of the French school of finite elements that started with Philippe Ciarlet and Pierre Arnaud Raviart, and that comes out of the French school of functional analysis and theoretical partial differential equations of Laurent Schwartz, Jacques Louis Lions, and their collaborators and students.

However, this book may differ from others because it focuses with more details on the treatment of boundary conditions, particularly non-homogeneous Dirichlet conditions, but also pure Neumann conditions, as well as nonlinear conditions of Signorini type.

As well, I tried to illustrate the practical implications of some theoretical results, such as well-posedness, but also error estimates. And, lastly, my hope is to target some colleagues or students within the finite element community who are interested

---

[1] See particularly their respective writings, among which [93] and [179]. The above quotation of Lothar Collatz is borrowed from the English translation of [93]. The preface of this book contains inspiring thoughts about mathematics and numerical analysis, as well as the preface of a chapter of Serguey Repin [227] that allowed me to discover these references.

in understanding better the mathematics, but who still have found no book they like to work with. Notably, the reader will find at the beginning of each chapter an outline of a few pages that explains the main ideas with the less possible formalism. The core of each chapter then provides more details for the reader eager to learn more.

Montevideo, Uruguay   Franz Chouly
May 2024

Acknowledgements

I thank warmly Chris Eder, Krishnakumar Pandurangan and the other members of the editorial team of Springer/Birkhauser for their help and support, as well as the following colleagues, for inspiring discussions and encouragements: Yves Achdou, Rodolfo Araya, Roland Becker, Faker Ben Belgacem, Stéphane Bordas, Juan Pablo Borthagaray, Marek Bucki, Raphaël Bulle, Erik Burman, Alfonso Caiazzo, Michel Duprez, Alexandre Ern, Mathieu Fabre, Miguel Fernández, Tom Gustafsson, Sergio Gutiérrez, Patrick Hild, Hao Huang, Alexei Lozinski, Ernesto Mordecki, Joaquin Mura, Jacques Ohayon, Pascal Omnes, Yohan Payan, Yves Renard, Pierre-Yves Rohan, and Mircea Sofonea.

Contents

1 Introduction 1

2 A Simple Boundary Value Problem 11

3 Low-Order Lagrange Finite Elements 51

4 The Standard Finite Element Method 63

5 Nitsche Finite Element Method 79

6 Nitsche for Signorini 99

7 A Posteriori Error Estimation 111

A More About Practical Implementation 125

Solutions 129

Bibliography 137

Index 151

Introduction

1

This book aims at presenting the finite element method from a mathematical viewpoint, with emphasis on the interest that mathematical analysis tools may have for practitioners. By practitioners I mean the vast community of colleagues who do practical simulation using the finite element technology to solve problems of interest in mathematics, physics, engineering, or other, but for whom it might not be obvious or relevant to get immersed into an in-depth study of the whole mathematical theory hidden behind finite elements. These practitioners can be engineers in the industry, scientists in applied laboratories, or mathematicians who are not specialists in numerical analysis but may have an interest in solving partial differential equations, for instance, for control problems or shape optimization. It is above all dedicated to students who want to enter into the subject while keeping one eye on each side, theoretical and applied.

The finite element method has a long story, with early contributions from both the "pure" mathematics and engineering communities (see, for instance, [147, 212] for an historical perspective), and, since the end of 1960s, many exchanges between the various communities of users and contributors, and many joint contributions. By the way, there exist many outstanding reference books and introductive books with various viewpoints, either they concern the mathematical foundation of the method or application aspects and implementation issues in large-scale finite element libraries.

Nonetheless, I feel there is still a gap to fill between the theory and the practice, and this is why I got motivated to write this book. First to provide a mathematically sound introduction to the method, but as simple as possible. Then to shed light on two specific topics. The first one is the detailed numerical treatment of boundary conditions, which is often skipped in most of the introductive material. The other one concerns the discretization error, precisely at the point where the theory meets the practice.

It is worth mentioning that a major source of inspiration for this book has been the French collection of books for master students, called "Mathématiques Appliquées pour la Maîtrise" and edited under the supervision of P.G. Ciarlet and J.L. Lions. I learnt a lot from this collection, notably with the book of P.A. Raviart and J.M. Thomas [222] that describes the finite element method, the book of H. Brezis focused on functional analysis and partial differential equations [60], and the book of G. Duvaut that introduces to continuum mechanics.

1.1   My Own Experience

I started to use the finite element method in 2002, during my PhD thesis. I was by this time in a laboratory of imaging, robotics, and biomechanics for clinical applications, where fascinating scientific and technological challenges had to be addressed. Particularly, I had to carry out simulations with a blackbox finite element solver, that provided, sometimes, plausible results. First I realized that colleagues more gifted than me in mechanics or physics used their own intuition a lot to get decent results. But then, I had a problem: I lack totally of physical intuition. In fact, sentences like "It is unphysical" have no clear meaning to me.

As a result, I tried to understand better the mathematics. I first went into the monograph of O.C. Zienkiewicz and R.L. Taylor [249], which I was told to be the most popular reference on the topic. From this first study, I got more information and I started to understand many things, but also many other things remained to be understood better. Notably because in such monographs oriented towards engineers, it does not appear clearly that a finite element solution approximates the solution to a boundary value problem, that is to say, of a partial differential equation completed with some specific boundary conditions. Emphasis is made above all on details relevant for implementation of finite elements for many mathematical models for mechanical engineers (elasticity, plates, etc).

Later on, after 2006, I had the chance to get into the study of books more focused on the mathematical theory, particularly the books of S. Brenner and R. Scott [59], and the (first) book of A. Ern and J.L. Guermond [125]. It helped me a lot to understand issues such as well-posedness at the discrete level and error estimates. At the same moment, I discovered FreeFEM++ [167] which was among the very first initiatives of finite element software that allowed to implement and test finite element methods directly with the mathematical formalism of weak formulations. This has been a great paradigm shift for me. Particularly, I realized then how much numerical analysis is important whenever one wants to design nonconventional finite element methods, such as stabilized or nonconforming methods, that work properly.

Some years later, after 2014, I had the opportunity to work with an industrial partner involved in a spinoff, and this took me back to practical problems. Notably, it allowed me to remember that many practitioners of the finite element method have to face difficulty in providing accurate solutions in realistic conditions, i.e. with somehow limited resources, and other difficulties to evaluate the reliability of their simulation pipeline. This topic was at the origin of a series of works, in which at first, we simply applied techniques of a posteriori error estimations that are rather standard within the community of theoretical numerical analysis, but largely unused by a large community of practitioners. This reminded me also to which point the scientific (sub)communities remained disconnected. I hope this book can be a small contribution to reconnect some practices of finite element simulations.

1.2   Overview of the Content and Chapters

In Chap. 2, we provide a simple mathematical model of a linear elliptic partial differential equation: Poisson's problem that represents, for instance, a diffusion process within a homogeneous continuous medium. Emphasis is made on the boundary conditions. First the setting for a (non-homogeneous) Dirichlet boundary condition and then for a pure Neumann boundary condition is detailed. These are indeed the two most representative cases. Since the weak formulation is the very basis of the discretization with finite elements, we explain it in detail as well as the strong-weak equivalence that allows to ensure that the problem approximated with finite elements is the same as the initial problem (this is not a trivial issue for some models). Also, we introduce here the notion of a well-posed problem: the mathematical model needs to have a solution and only one, and this solution should depend continuously on the data. This notion will play an important role in the mathematical analysis of the discretization. For each setting, we describe the strong form and the weak form of the equations, as well as elementary results of well-posedness. This is the ground before discretization.

In Chap. 3, we build the vector space of lowest-order Lagrange finite elements. Indeed, a second important idea of the finite element method is to approximate the function space in which the solution to the weak form is supposed to live. The first step is to mesh the domain, which means to split it into small cells of simple shape. Then we introduce the space of piecewise linear Lagrange finite elements on simplicial meshes, which means that we approximate any function using first-order polynomials on each simplicial cell. We provide finally a result of polynomial approximation that will be useful to derive some error bounds.

Chapter 4 glues the two previous ones and presents the standard way to approximate Poisson's problem as described in Chap. 2. To summarize, the technology relies on (1) writing the mathematical model in weak form and (2) approximating the vector space where the solution lives, using the finite-dimensional vector space designed in Chap. 3 (Galerkin method). This allows to transform the original boundary value problem and to get at the end a simple linear system to build and solve. At this point, a computer will do this job very efficiently. We provide results of well-posedness at the discrete level. Practically, this will ensure that the matrix involved in the final linear system is always invertible. The next question addressed is those of a priori error bounds. Particularly, they guarantee that the approximation error vanishes when the mesh is refined, with the expected (optimal) rate. It is an important property that any method needs to satisfy. We treat the case of both Dirichlet and Neumann boundary conditions.

Chapter 5 details the Nitsche method to impose essential boundary conditions, as a prototype of a non-standard method. The interest of the mathematical analysis will appear more clearly here, since it helps in the design of a method with the right numerical performance. We detail the discrete well-posedness and how a priori error bounds can be obtained.

Chapter 6 presents how the standard and Nitsche methods can be extended to the Signorini problem. The Signorini problem involves nonlinear boundary conditions and is a prototype for modelling contact in computational mechanics. As explained previously, we demonstrate how to derive the methods and present their mathematical properties of well-posedness and convergence.

Chapter 7 focuses on the discretization error in some situations of practical interest. We enter into the universe of a posteriori error bounds that allow us to assess the magnitude of the discretization error and refine the mesh where the local error is the largest. We present a common a posteriori error estimator (explicit residual based estimator) as well as techniques for goal-oriented error estimation, to assess the accuracy for a specific quantity targeted by the user.

Finally, Appendix A presents an example of the implementation of the Nitsche method in a high-level finite element library (scikit-fem).

1.3   Prerequisites and Advices

The reader is expected to have basic undergraduate notions in mathematics, and it should, hopefully, be enough. More advanced notions are introduced along with their motivation in each chapter. Reading Chaps. 2–4 will help the reader to be familiar with all the basic aspects of the finite element method. The further chapters are independent and can be studied separately. An outline at the beginning of each chapter attempts at presenting the key ideas with the less possible mathematical sophistications. Then we try to detail as much as possible the different concepts and statements.

1.4   Basic Notions About Functions and Distributions

Here we simply recall the strictly necessary notions about functions, differential operators, and distributions that we need in this book. For more details, the interested reader may refer, for instance to [232]; see also [5, 60, 86, 127, 148, 233, 241].

1.4.1   Functions and Differential Operators

For $d \geq 1$, let ω be a subset of $\mathbb{R}^d$. The notation $x = (x_1, x_2, \ldots, x_d)$ stands for a generic point in ω. Let $\alpha = (\alpha_1, \alpha_2, \ldots, \alpha_d) \in \mathbb{N}^d$ be a multi-index of order $|\alpha| = \alpha_1 + \alpha_2 + \ldots + \alpha_d \in \mathbb{N}$. For a regular enough function $\psi : \omega \to \mathbb{R}$, the partial derivative of ψ with respect to the multi-index α is

$$D^\alpha \psi := \frac{\partial^{|\alpha|} \psi}{\partial x_1^{\alpha_1} \partial x_2^{\alpha_2} \dots \partial x_d^{\alpha_d}}.$$

For instance, in dimension two ($d = 2$), the three partial derivatives of order two are denoted by:

$$D^{(2,0)}\psi = \frac{\partial^2 \psi}{\partial x_1^2}, \quad D^{(1,1)}\psi = \frac{\partial^2 \psi}{\partial x_1 \partial x_2}, \quad D^{(0,2)}\psi = \frac{\partial^2 \psi}{\partial x_2^2}.$$

The gradient of ψ is a vectorial field on ω with values in $\mathbb{R}^d$ defined as follows:

$$\nabla \psi := \left(\frac{\partial \psi}{\partial x_1}, \frac{\partial \psi}{\partial x_2}, \dots, \frac{\partial \psi}{\partial x_d} \right).$$

For a regular enough vector-valued function $\psi : \omega \to \mathbb{R}^d$, using the notation $\psi := (\psi_1, \psi_2, \dots, \psi_d)$ for its components, we define its divergence as

$$\operatorname{div} \psi := \frac{\partial \psi_1}{\partial x_1} + \frac{\partial \psi_2}{\partial x_2} + \dots + \frac{\partial \psi_d}{\partial x_d}.$$

The set of continuous real-valued functions in ω is denoted by $\mathscr{C}(\omega)$. In the same way, the set of real-valued functions in ω that admit continuous Gâteaux-derivatives up to order k ($k \geq 0$) is denoted by $\mathscr{C}^k(\omega)$ (and we recover $\mathscr{C}^0(\omega) = \mathscr{C}(\omega)$). The set of real-valued functions in ω that are infinitely Gâteaux-differentiable is noted $\mathscr{C}^\infty(\omega)$, i.e.

$$\mathscr{C}^\infty(\omega) = \bigcap_{k \geq 0} \mathscr{C}^k(\omega).$$

For vector-valued functions, we will use the notation $\mathscr{C}(\omega; \mathbb{R}^d)$ for continuous vector-valued functions defined in ω and that take their values in $\mathbb{R}^d$. In the same manner, we note $\mathscr{C}^k(\omega; \mathbb{R}^d)$, respectively $\mathscr{C}^\infty(\omega; \mathbb{R}^d)$, for functions with continuous derivatives up to order k, respectively, infinitely differentiable.

1.4.2   Test Functions

Let now Ω be an open subset of $\mathbb{R}^d$. For any function $f : \Omega \to \mathbb{R}$, the support of f is the closure of the set where f does not vanish:

$$\operatorname{supp} f = \overline{\{x \in \Omega \mid f(x) \neq 0\}} \subset \mathbb{R}^d.$$

We say that f has a compact support if there exists a compact set $K \subset \Omega$ such that $\operatorname{supp} f \subset K$. Notably, if f has compact support, it vanishes in a neighbourhood of the boundary $\partial \Omega$ of the domain Ω. The set of test functions, denoted by $\mathscr{D}(\Omega)$, is the set of infinitely Gâteaux-differentiable functions with a compact support, i.e.,

$$\mathscr{D}(\Omega) := \{f \in \mathscr{C}^{\infty}(\Omega) \mid \exists K \subset \Omega, K \text{ compact, supp } f \subset K\}.$$

For explicit examples of test functions with useful properties of localization, for instance, see [86, 148, 232, 233]. We endow $\mathscr{D}(\Omega)$ with the following notion of convergence:

Definition 1.1 A sequence of functions $(\varphi_n)_{n \in \mathbb{N}}$ tends to 0 in $\mathscr{D}(\Omega)$ if:

1. There exists a compact set $K \subset \Omega$ such that for all $n \in \mathbb{N}$:

$$\text{supp } \varphi_n \subset K.$$

2. For any multi-index α, the function $D^{\alpha} \varphi_n$ tends to 0 uniformly in K:

$$\max_{x \in \Omega} |D^{\alpha} \varphi_n(x)| \xrightarrow{n \to +\infty} 0.$$

We will use the following notation for this convergence:

$$\varphi_n \xrightarrow{\mathscr{D}(\Omega)} 0.$$

For $\varphi \in \mathscr{D}(\Omega)$, we say that $\varphi_n \xrightarrow{\mathscr{D}(\Omega)} \varphi$ if and only if $(\varphi_n - \varphi) \xrightarrow{\mathscr{D}(\Omega)} 0$.

Finally, the vector space of vector-valued test functions defined in Ω and with values in $\mathbb{R}^d$ is denoted by $\mathscr{D}(\Omega; \mathbb{R}^d)$.

1.4.3 Distributions

Let us now recall the notion of distribution:

Definition 1.2 A distribution T is a continuous linear form on $\mathscr{D}(\Omega)$, which means that for any sequence $(\varphi_n)_{n \in \mathbb{N}}$ of test functions such that

$$\varphi_n \xrightarrow{\mathscr{D}(\Omega)} 0,$$

there holds:

$$\langle T, \varphi_n \rangle \xrightarrow{n \to +\infty} 0,$$

where $\langle T, \varphi \rangle$ denotes the duality pairing between $\mathscr{D}(\Omega)$ and its topological dual, denoted as $\mathscr{D}'(\Omega)$. The vector space of distributions is thus denoted as $\mathscr{D}'(\Omega)$.

Distributions always admit derivatives of any order, if we define the derivative in the following sense:

Definition 1.3 Let $T \in \mathscr{D}'(\Omega)$ be a distribution and α be a multi-index. The partial derivative of T with respect to the multi-index α is denoted as $D^\alpha T \in \mathscr{D}'(\Omega)$ and defined by

$$\langle D^\alpha T, \varphi \rangle = (-1)^{|\alpha|} \langle T, D^\alpha \varphi \rangle$$

for all $\varphi \in \mathscr{D}(\Omega)$.

One readily checks that this definition makes sense and that $D^\alpha T$ belongs indeed to $\mathscr{D}'(\Omega)$. This definition is inspired by integration by parts. The partial derivative of a distribution T is called a distributional derivative.

We can also define vector-valued distributions as follows: $T = (T_1, \ldots, T_d) \in \mathscr{D}'(\Omega; \mathbb{R}^d)$ is a vector-valued distribution if each of its components $T_i, i = 1, \cdots, d$, is a (scalar) distribution ($T_i \in \mathscr{D}'(\Omega)$), and there holds for every vector-valued test function $\phi = (\varphi_1, \ldots, \varphi_d) \in \mathscr{D}(\Omega; \mathbb{R}^d)$:

$$\langle T, \phi \rangle = \sum_{i=1}^{d} \langle T_i, \varphi_i \rangle.$$

1.4.4 Regular Distributions

A large class of functions can be viewed as distributions. This is the class of locally integrable functions. The vector space of locally integrable functions on Ω is defined as follows:

$$L^1_{\text{loc}}(\Omega) := \left\{ f : \Omega \to \mathbb{R} \,\middle|\, \int_K |f(x)| \mathrm{d}x < +\infty, \, \forall K \subset \Omega, \, K \text{ compact} \right\}.$$

To any function $f \in L^1_{\text{loc}}(\Omega)$, we can associate a distribution $T_f \in \mathscr{D}'(\Omega)$ using the following definition:

$$\langle T_f, \varphi \rangle = \int_\Omega f(x)\varphi(x)\, \mathrm{d}x, \tag{1.1}$$

for all $\varphi \in \mathscr{D}(\Omega)$. Since $f \in L^1_{\text{loc}}(\Omega)$, and φ is continuous and has compact support, the above definition is meaningful.

For locally integrable functions, there holds the following variational lemma, whose importance is considerable in applied analysis:

Lemma 1.1 *Let $f \in L^1_{loc}(\Omega)$ such that*

$$\int_\Omega f(x)\varphi(x)\,dx = 0$$

for all $\varphi \in \mathscr{D}(\Omega)$. Then, $f = 0$ almost everywhere in Ω.

For a proof based on the Lebesgue Differentiation Theorem, see, for instance, [232, Proposition 1.1]. This lemma means in some sense that the space of test functions is rich enough to characterize any locally integrable function. The first important consequence of the above lemma is the injectivity of the application

$$L^1_{loc}(\Omega) \ni f \mapsto T_f \in \mathscr{D}'(\Omega),$$

so that $L^1_{loc}(\Omega)$ can be identified with a subspace of $\mathscr{D}'(\Omega)$. This particular subspace is called the subspace of regular distributions. As usual with distributions, we will use the notation f instead of T_f for the regular distribution associated to a locally integrable function f.

1.4.5   The Heaviside and the Dirac Delta Functions

We end this section with an example that will have great importance in the sequel. To simplify, we take $\Omega = (-1; 1)$ and introduce the Heaviside function H, more precisely, its restriction on $(-1; 1)$:

$$H : x \mapsto \begin{cases} 0 & \text{if } x < 0, \\ 1 & \text{if } x > 0. \end{cases} \tag{1.2}$$

Figure 1.1 depicts its graphical representation.

Clearly, we have $H \in L^1_{loc}(\Omega)$, and its distributional derivative is the Dirac delta function

$$H' = \delta. \tag{1.3}$$

We recall that $\delta \in \mathscr{D}'(\Omega)$ is defined in a unique manner by

$$\langle \delta, \varphi \rangle = \varphi(0), \tag{1.4}$$

for $\varphi \in \mathscr{D}(\Omega)$. Moreover, there holds $\delta \notin L^1_{loc}(\Omega)$, which means that δ is not a regular distribution (and implies that $L^1_{loc}(\Omega)$ is a proper subspace of $\mathscr{D}'(\Omega)$). Though it is not obvious to represent graphically distributions that are not standard

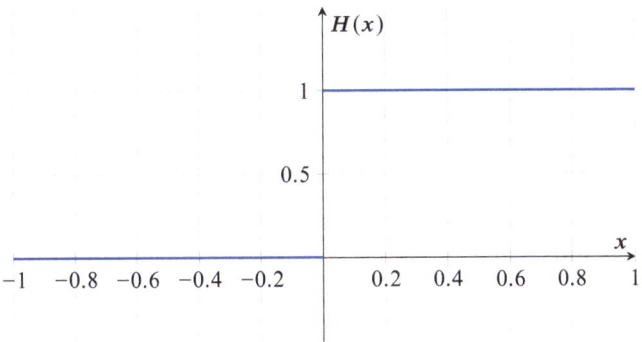

Fig. 1.1 Heaviside step function $H(x)$

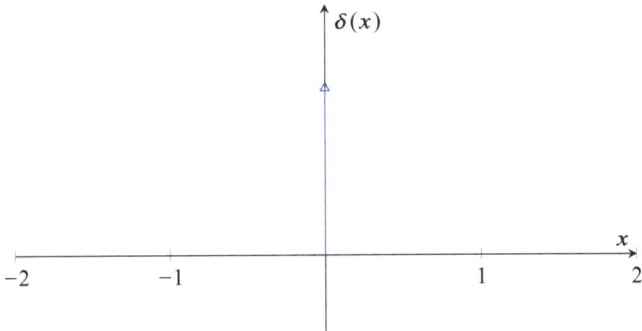

Fig. 1.2 Dirac delta function $\delta(x)$

functions, a common representation of δ as a spike at the origin is depicted in Fig. 1.2.

1.5 Further Comments

The theory of distributions is attributed to L. Schwartz, who spent the years 1945–1950 in elaborating this theory. He was awarded the Fields Medal in 1950 due to this breakthrough. His book [233] remains a classical reference. It has been motivated to generalize the concepts of function and derivative because of some issues related to the solutions of partial differential equations and to the Fourier transform, among others. Between the important contributions that led to the theory of distributions were pioneering works of J. Fourier (1822), G.R. Kirchhoff (1882), and O. Heaviside (1898) that made use of distributions before the rigorous theory was achieved. There were also contributions of S. Bochner (1932) and S. Sobolev (1935) towards its formalization. Particularly, S. Sobolev is considered at the origin of the notion of weak derivative, which is a special case of the distributional derivative. For a historical view on distributions, one can refer to the book of J. Lutzen [195], or also [233, 241]. Introductive material to distributions can be found in the book of C. Gasquet and P. Witomski [148]. For a presentation more oriented to partial differential equations, one can refer, for instance, to the corresponding chapters of L. Tartar [241].

A Simple Boundary Value Problem

2

This chapter details a simple diffusion model as a prototype of a boundary value problem based on a linear, scalar, stationnary partial differential equation. This partial differential equation allows to recover the behavior of a solution function on a given spatial domain, and this solution function represents generally a physical quantity (here physical may be understood in the broadest sense). At the boundary of the domain, other conditions need to be stated on the solution function so that the problem becomes mathematically meaningful. The combination of a partial differential equation and some boundary conditions is called a boundary value problem.

2.1 Outline

We start from practice in numerical simulation and emphasize there is a pipeline between the real-life system that motivated the simulation and the numerical prediction provided by the computer. We focus on the first step of this pipeline, which is the mathematical model. The notion of boundary value problem is illustrated with Poisson's problem, complemented with either a Dirichlet boundary condition or a Neumann boundary condition. We point out some mathematical properties that are worth noting and we would like to preserve them later on with a numerical approximation.

2.1.1 Mathematical Models in Numerical Simulation

When a numerical simulation is carried out, there is properly speaking no model inside the computer. The computer is only capable of providing you a sequence of (floating point arithmetic) numbers as a prediction of the behavior of the real-

© The Author(s), under exclusive license to Springer Nature Switzerland AG 2025
F. Chouly, *Finite Element Approximation of Boundary Value Problems*,
Compact Textbooks in Mathematics, https://doi.org/10.1007/978-3-031-72530-2_2

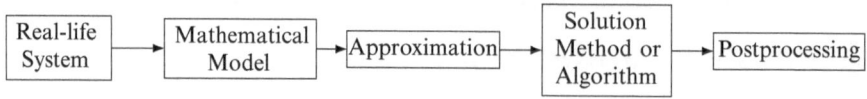

Fig. 2.1 The above pipeline depicts the global process behind a numerical simulation

life system you are interested in. These numbers have been obtained after many arithmetic operations, starting from other numbers as input.

On the other hand, numerical simulations are used to predict the behavior of some real-life systems in physics, engineering, or other scientific disciplines (more or less) connected to the real world.

So between both, there is a pipeline. The first step of this pipeline is mathematical modelling. The other steps are depicted below (Fig. 2.1).

In the second step, the mathematical model is approximated into something that can be processed by a computer. In the third step, an algorithm, implemented in any programming language, runs inside a given computer and provides a solution, expected to be similar to the solution of the original mathematical model. Finally, in the fourth step, this solution is postprocessed, in order to deliver some numbers useful to the user.

We will get back to the other steps in the further chapters. Let us go back to the first step to emphasize, that, in fact, when a numerical simulation is carried out on a computer, there is most of the time an exact mathematical model behind, that is, a set of equations and unknowns to these equations. If you have some interest in understanding better the mathematics, the first step is to get more insight into this mathematical model, at least to know with accuracy which equations are approximated and solved.

The mathematical model can be obtained from a combination of basic physical principles (conservation of mass, momentum, etc.) and some general empirical laws of physics, biology, material science, etc. It allows to predict the behavior, in time, in space, or both, of the real-life system, with some degree of generality and of reliability. The differences there are between the predictions from the mathematical model and between the measurements and observations picked from the real-life systems are called modelling errors. This is a vast, important, and complicated issue, and we will not cover it extensively in this book (unfortunately); see, however, the end of Chap. 7 for a few extra comments. Let us now turn to a simple set of equations, as a prototype of a mathematical model for the finite element method.

2.1.2 Poisson's Problem as an Example of Boundary Value Problem

In this book, we consider a very special class of mathematical problems, which are boundary value problems. These are made of two blocks: (1) a partial differential equation on a given, bounded, domain; (2) some additional conditions on the boundary. We provide an emblematic example of boundary value problem with all the features suited for a discretization with the finite element method. This is Poisson's problem below, with first the Laplace (diffusion) equation as a partial differential equation:

$$- \mu \Delta u = f \tag{2.1}$$

in a domain Ω of dimension one, two, or three. Above, u is the solution to be sought (a function in Ω), f is a source term, μ is a physical coefficient (the diffusion coefficient), and Δ is the Laplace operator. As for ordinary differential equations, Eq. (2.1) alone is not enough to predict the behavior of a given real-life system, because it undetermines u.

Other conditions need to be introduced, and it is usual to prescribe a relationship that u verifies on the boundary $\Gamma := \partial \Omega$ of the domain Ω. This is called a boundary condition. Many possibilities exist. For instance, we can prescribe directly the value of u on the boundary, which is called a Dirichlet boundary condition. So we set

$$u = g \tag{2.2}$$

on Γ, where the unknown u is directly fixed and equal to a given function g. Such a condition is called an *essential* boundary condition. This terminology comes from the weak formulation: an essential boundary condition is incorporated directly within the function space where the solution is sought. The combination of (2.1) and (2.2) forms a first example of a boundary value problem.

2.1.3 Weak Formulation

The first ingredient for finite element approximation is to transform a boundary value problem such as (2.1)–(2.2) into a weak, or variational, formulation. This is made possible thanks to the multiplication of the equation by an arbitrary test function and then using the Green formula. Poisson's problem in weak form reads:

Find $u \in V$ that satisfies $u|_\Gamma = g$ on Γ and

$$\mu \int_\Omega \nabla u \cdot \nabla v = \int_\Omega f v \quad \text{for all } v \in V \text{ with } v|_\Gamma = 0, \tag{2.3}$$

where ∇ is the distributional gradient, $|_\Gamma$ denotes the restriction of a function on the boundary Γ, and V is a vector space of functions. This space V has a special mathematical structure, as a particular class of Hilbert space, and is called a Sobolev space. In addition, the weak formulation (2.3) can be obtained from the energy functional

$$\mathcal{J}(v) := \frac{1}{2} \mu \int_\Omega \nabla v \cdot \nabla v - \int_\Omega f v$$

defined for $v \in V$. Indeed, it suffices to explicit the first-order condition, $\mathcal{J}'(u, \cdot) = 0$, for its minimization under the equality constraint $v|_\Gamma = g$. Unfortunately, this variational viewpoint is not always possible and some boundary problems cannot be derived from energy minimization. It is the case, for instance, of many models in fluid mechanics. However, these models may have a weak formulation that makes possible finite element approximation; see e.g. [232] and [127–129].

2.1.4 Well-Posedness

A fundamental issue before going into numerical approximation is to ensure that Poisson's problem (2.3) is *well-posed*, using the definition coming from Hadamard. This has important implications at the numerical level, particularly it ensures that:

- The solution u to (2.3) exists, which is a minimal requirement. When there is no solution, the mathematical model needs to be revisited, generally because it has missed something important from the underlying physical laws and empirical conditions.
- The solution u to (2.3) is unique. It should not be a requirement of the model since many physically relevant models can have multiple (physically relevant) solutions, but this needs to be known prior to doing numerics. Indeed, if no special treatment such as bifurcation tracking is done, the numerical method may select "randomly" one solution among the others.
- The solution depends continuously on the data. In our case, where the problem is linear, this means that small f and large μ should imply a small solution u, in a sense that will be precise. This ensures a nice property of numerical stability for a large class of reasonable discretization techniques.

This well-posedness issue can be solved elegantly thanks to elementary notions of functional analysis:

1. Thanks to the (homogeneous) Dirichlet boundary condition on the test functions v, we can make use of a *Poincaré-Friedrichs* inequality to prove that the bilinear form

$$(v, w) \mapsto \mu \int_\Omega \nabla v \cdot \nabla w \in \mathbb{R}$$

   defines an inner product on $V_0 := \{v \in V \mid v|_\Gamma = 0\}$.
2. Since V is a Hilbert space and V_0 a closed subspace of V, the Riesz-Fréchet Representation Theorem is used to ensure the existence of a solution u.

The main difficulty comes from the non-homogeneous Dirichlet condition $u = g$ on the boundary Γ and is treated here using a *lifting* of g from the boundary to the domain, in other terms a function u_g in V that is equal to g when restricted to the boundary. The construction of this lifting is a delicate issue. The continuous dependence is finally obtained by estimations that provide a bound for the solution u that depends on f and μ. Finally, it remains important to prove the equivalence between the strong (original) form of Poisson's problem and the weak form. This is almost a formality here, but knowing how to do this allows one to avoid errors when one is faced with more complex formulations.

2.1.5 Neumann Boundary Condition

Alternatively, a (pure) Neumann condition

$$\mu \nabla u \cdot n = h, \tag{2.4}$$

can be imposed instead of (2.2) to complement (2.1). Above, the value of the flux $\mu \nabla u \cdot n$ is prescribed on the boundary Γ and is equal to a flux h (n is the outward unit normal to the boundary Γ). Poisson's problem can be once again transformed into a weak form. Establishing its well-posedness requires special care, since the source term f and the boundary flux term h need to satisfy a compatibility condition to ensure the existence of solutions, and the solution is in fact an equivalence class.

2.1.6 Regular and Singular Solutions

We end the chapter with a few considerations about the regularity of the solution. If the domain Ω has a smooth boundary, or for a convex polygon in two dimensions, and if the source term f is smooth, the solution u is also smooth.

However, in many practical situations, singularities may appear. For instance, for nonconvex polygons in two dimensions, especially for cracks, singularities appear at reentrant corners, especially at the crack tip. As we will see in the next chapters, the standard finite element method is not at its best in such situations. Particularly, these singularities can deteriorate the performance or the accuracy of the simulation output. There are many remedies to this illness, and we will present one at the end of this book; see Chap. 7.

2.2 A Short Preliminary

Before describing in detail the mathematical model (the boundary value problem in our case), we need to recall a few notions about the domain and related assumptions. We will insist on the specific class of Lipschitz domains that play a central role in the theory of elliptic partial differential equations and their approximation. We pursue this preliminary by stating the Divergence Theorem and a simple Green formula. We end with the statement of one important result to establish the well-posedness of various problems under consideration: the Riesz-Fréchet Representation Theorem.

2.2.1   The Domain and Its Boundary

Partial differential equations describe the variations of a quantity over a spatial domain and are generally built using conservation laws coming from physics that involve the various partial derivatives of this quantity. So, first, we need to be accurate about what we call a domain and its boundary: we consider first that a domain Ω is an open subset of $\mathbb{R}^d$, with $d = 1, 2, 3$. The boundary of Ω is denoted by $\Gamma(:= \partial\Omega)$. Other assumptions about the geometrical or topological properties of Ω will be made when needed. First, in this presentation, we will restrict to the large class of Lipschitz domains. We follow the presentation of [232] and recall only briefly the main concepts.

2.2.2   Lipschitz Domains

First, we introduce the following notion:

Definition 2.1 Let $\mho$ be an open subset of $\mathbb{R}^d$. The application $F : \mho \to \Omega_F := F(\mho) \subset \mathbb{R}^d$ is a bi-Lipschitz homeomorphism if F is invertible from $\mho$ to Ω_F and if both F and F^{-1} are Lipschitz. This means there exists positive constants c and C such that for every x and y in $\mho$, there holds

$$c\,|x - y| \le |F(x) - F(y)| \le C\,|x - y|.$$

Let $B_{d-1}(0, 1)$ be the open unit ball in $\mathbb{R}^{d-1}$. We now take $\mho := B_{d-1}(0, 1) \times (-1, 1)$ as a reference cylinder in $\mathbb{R}^d$ and define then the following subdomains of $\mho$:

$$\mho^+ := B_{d-1}(0, 1) \times (0, 1),$$
$$\mho^- := B_{d-1}(0, 1) \times (-1, 0),$$
$$\Sigma := B_{d-1}(0, 1) \times \{0\}.$$

The above notions and notations allow to define a Lipschitz domain, as, roughly speaking, a collection of charts that are Lipschitz deformations of $\mho$ that preserve the above structure (a boundary that separates an interior from an exterior). We consider the case of a bounded domain for the sake of simplicity.

Definition 2.2 Let Ω be a bounded open subset in $\mathbb{R}^d$. We say that Ω is a Lipschitz domain if there exists a finite collection of charts (F_j) and open sets (Ω_j), $j = 1, \ldots, J$, with

$$F_j : \mho \to \Omega_j := F_j(\mho) \subset \mathbb{R}^d,$$

that are bi-Lipschitz homeomorphisms (for $j = 1, \ldots, J$) and that verify the
following relationships:

$$\Omega_j^+ := F_j(\mho^+) = \Omega_j \cap (\overline{\Omega})^c,$$

$$\Omega_j^- := F_j(\mho^-) = \Omega_j \cap \Omega,$$

$$\Gamma_j := F_j(\Sigma) = \Omega_j \cap \Gamma, \quad \forall j = 1, \ldots, J.$$

and such that

$$\bigcup_{j=1}^{J} \Gamma_j = \Gamma.$$

We also introduce Ω_0 as a subset of Ω such that

$$\Omega \subset \bigcup_{j=0}^{J} \Omega_j.$$

The construction associated with Definition 2.2 is illustrated in Fig. 2.2.

Remark 2.1 Interestingly, every domain Ω that is both convex and bounded can be
proved to be Lipschitz; see e.g. [103, Lemma 2.3] and [156].

2.2.3 Examples and Counterexamples of Lipschitz Domains

First, all bounded domains with very smooth boundaries such as circles, ellipses,
etc., are Lipschitz domains More importantly, in two dimensions, all polygons with
interior angles strictly lower than 2π are Lipschitz. In three dimensions also, a large
variety of polyhedra are Lipschitz. See Fig. 2.3 for some illustrations.

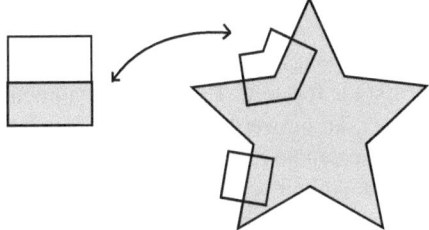

Fig. 2.2 Covering of a Lipschitz domain Ω (in grey, on the right). The boundary Γ of Ω is such
that it can be covered by a finite union of sets Ω_j that are images of $\mho$ (on the left) by a bi-Lipschitz
homeomorphism. Moreover, this covering preserves the partition of $\mho$ into $\mho^-$ (gray), Σ (black),
and $\mho^+$ (white)

Fig. 2.3 All the above
domains Ω (gray) in $\mathbb{R}^2$ are
Lipschitz domains

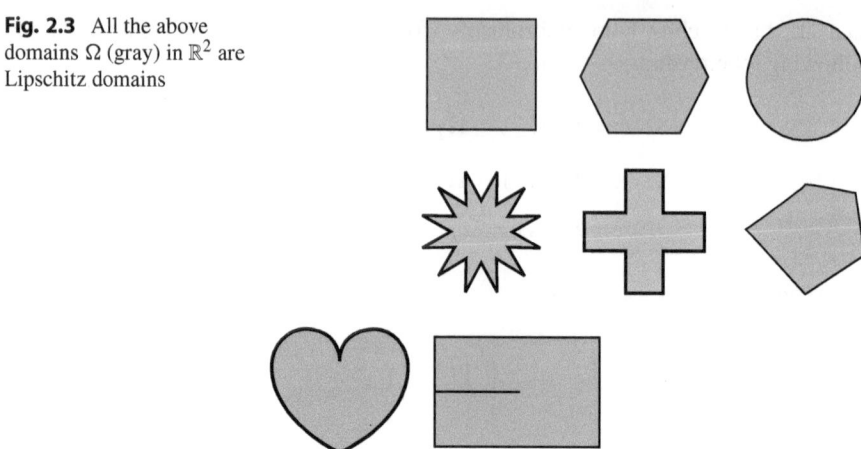

Fig. 2.4 The two above domains Ω (grey) in $\mathbb{R}^2$ are not Lipschitz domains. Unfortunately, the
nice heart on the left is not Lipschitz, and this is the case for all the domains with cusps. Also, the
crack domain on the right is not Lipschitz

However, domains with cracks or tips are not Lipschitz, as well as domains
with a cusp. There is indeed an impossibility near some specific points (the cusp
and the crack tip) to check the Definition 2.2. Unfortunately, these domains are
not rare in computational mechanics since they may be useful to model fracture or
faults, for instance, which is very useful in structural engineering. See Fig. 2.4 for
some illustrations. Some other illustrations of examples and counterexamples can
be found in e.g. [156, 198, 232].

2.2.4 Normal to the Boundary

First, we recall here Rademarcher's Theorem (see e.g. [133, 3.1.6, p.216]):

Theorem 2.1 *Let* $f : \mathbb{R}^d \to \mathbb{R}^m$, $d, m \geq 1$, *be a Lipschitz function. Then* f *is
Gâteaux-differentiable almost everywhere in* $\mathbb{R}^d$.

This theorem allows to define a surface measure on a Lipschitz boundary Γ. This
measure is denoted by $ds(x)$, for $x \in \Gamma$ (x will be sometimes omitted when it
is not necessary). Moreover, an outward unit normal vector can be defined almost
everywhere on Γ as well. We denote it by $n(x)$, for every point $x \in \Gamma$ such that this
vector exists. Detailed (and explicit) construction of this surface measure and this
normal vector has been done carefully in e.g. [198] or [232]. Of course, for polytopal
domains i.e. polygons in two dimensions and polyhedra in three dimensions, which
will be mostly our domains of interest in the next chapters; the construction of this
surface measure $ds(x)$ and outward unit normal $n(x)$ is of no difficulty.

2.2.5   The Divergence Theorem and Green Formula

Let us now recall the Divergence Theorem (or Gauss Theorem; see e.g. [198, Theorem 3.34] for the proof):

Proposition 2.1 (Gauss) *Let Ω be a Lipschitz domain in $\mathbb{R}^d$, of boundary $\Gamma := \partial\Omega$, and $\psi \in \mathscr{C}^1(\overline{\Omega}; \mathbb{R}^d)$ be a vector field. Then, there holds*

$$\int_\Omega \operatorname{div} \psi = \int_\Gamma \psi \cdot n, \tag{2.5}$$

where n is the outward unit normal to the boundary Γ.

We define the Laplace operator Δ as follows:

$$\Delta u := \operatorname{div} (\nabla u) \left(= \frac{\partial^2 u}{\partial x_1^2} + \frac{\partial^2 u}{\partial x_2^2} + \ldots + \frac{\partial^2 u}{\partial x_d^2} \right),$$

where the differential operators and derivatives can be understood at first in the classical sense. As a direct consequence of the Gauss Theorem, we can state a Green formula:

Proposition 2.2 *Let Ω be a Lipschitz domain in $\mathbb{R}^d$, of boundary $\Gamma := \partial\Omega$. Let $u \in \mathscr{C}^2(\overline{\Omega})$ and $v \in \mathscr{C}^1(\overline{\Omega})$. Then, there holds*

$$-\int_\Omega (\Delta u)v = \int_\Omega \nabla u \cdot \nabla v - \int_\Gamma (\nabla u \cdot n)v, \tag{2.6}$$

where n is the outward unit normal to the boundary Γ.

Proof Thanks to our assumptions on u and v, there holds

$$v\nabla u \in \mathscr{C}^1(\overline{\Omega}; \mathbb{R}^d)$$

and we apply the Divergence Theorem choosing $\psi = v\nabla u$. We get

$$\int_\Omega \operatorname{div} (v\nabla u) = \int_\Gamma (v\nabla u) \cdot n.$$

Then we write

$$\operatorname{div} (v\nabla u) = v\operatorname{div} (\nabla u) + \nabla u \cdot \nabla v = v\Delta u + \nabla u \cdot \nabla v$$

and this ends the proof.   □

We introduce the notation

$$\partial_n u := \nabla u \cdot n \tag{2.7}$$

for u a regular enough function on Γ and n the normal to Γ. This notation will be useful in the sequel.

Remark 2.2 For $d = 1$, we can take, for instance, $\Omega = (a, b)$, with $a, b \in \mathbb{R}$, and Green formula (2.6) reduces to the well-known integration-by-parts formula:

$$-\int_a^b \frac{d^2 u}{dx^2} v = \int_a^b \frac{du}{dx} \frac{dv}{dx} - \frac{du}{dx}(b)v(b) + \frac{du}{dx}(a)v(a).$$

2.2.6 The Riesz-Fréchet Representation Theorem

Let V be a real Hilbert space associated with inner product $(\cdot, \cdot)_V$ and norm $\| \cdot \|_V$. Let us denote by V' its topological dual that contains all the continuous (bounded) linear forms on V, endowed with the dual norm

$$\|l\|_{V'} := \sup_{v \in V, \|v\|_V = 1} l(v).$$

The dual space V' is also a Hilbert space. All the existing results of this chapter rely on the following theorem of functional analysis, whose proof can be found in [60, Theorem 5.5] or [232, Theorem 4.2]:

Theorem 2.2 (Riesz-Fréchet Representation Theorem) *Let V be a Hilbert space and V' its topological dual. Then for each l $\in$ V', there exists one unique $v_l \in$ V such that*

$$l(w) = (v_l, w)_V \quad \forall w \in V.$$

Moreover, there holds $\|v_l\|_V = \|l\|_{V'}$.

Remark 2.3 As a result, a Hilbert space V is isometrically isomorphic to its dual space V'. The above result is in some sense the most simple to prove some existence results for elliptic partial differential equations, but is limited to a special class of problems. For more sophisticated partial differential equations, more sophisticated results of functional analysis are needed [60, 125, 189, 232].

2.3   Poisson's Problem with a Dirichlet Boundary Condition

We present here in detail Poisson's problem with non-homogeneous Dirichlet boundary condition and its mathematical analysis: we prove strong-weak equivalence and well-posedness of this problem. In addition, we will recall its formulation as a minimization problem. Furthermore, we will emphasize the role of the lifting operator that allows to reformulate this problem to an equivalent, simpler, problem with homogeneous Dirichlet boundary conditions.

2.3.1   From the Strong Form to a Mock Weak Form

So, let a domain Ω be an open subset of $\mathbb{R}^d$ ($d = 1, 2, 3$), supposed, for the sake of simplicity, Lipschitz and bounded. We introduce Poisson's problem, on the aforementioned domain Ω and associated with the following data: a physical parameter $\mu \in \mathbb{R}$ (the diffusion coefficient), with $\mu > 0$, a source term $f : \Omega \to \mathbb{R}$, which is a prescribed scalar field in the domain and a boundary term $g : \Gamma \to \mathbb{R}$, which is in the same way a prescribed scalar field on the boundary. Then Poisson's problem with non-homogeneous Dirichlet boundary condition is the following boundary value problem:

Find $u : \Omega \to \mathbb{R}$ that satisfies :

$$\begin{cases} -\mu \Delta u = f & \text{in } \Omega, \quad (i) \\ \quad\;\; u = g & \text{on } \Gamma. \quad (ii) \end{cases} \tag{2.8}$$

Physically speaking, the above equation can model many phenomena related to a simple diffusion process. For instance, it can be used to compute the distribution of temperature in a homogeneous medium when a stationary state has been reached; see, for instance, [9]. Alternatively, in two dimensions, the unknown u can represent the vertical displacement of a thin membrane that occupies the domain Ω, in response to a density of surface forces f. The Dirichlet condition $u = g$ on the boundary Γ means that we enforce a known displacement g on the boundary.

Now let us introduce the two following auxilliary problems:

Find $u_0 : \Omega \to \mathbb{R}$ that satisfies:

$$-\mu \Delta u_0 = f \quad \text{in } \Omega, \qquad u_0 = 0 \quad \text{on } \Gamma, \tag{2.9}$$

and

Find $u_g : \Omega \to \mathbb{R}$ that satisfies:

$$-\mu \Delta u_g = 0 \quad \text{in } \Omega, \qquad u_g = g \quad \text{on } \Gamma. \tag{2.10}$$

Here we separated the contributions of the source term f and the boundary condition g. Note now that if u_0 is a solution to the Problem (2.9) with homogeneous boundary conditions, and if u_g is a solution to the Problem (2.10) with zero source term, then $u = u_0 + u_g$ solves the original Poisson's problem (2.8), because both the differential operator and the boundary condition are linear. This is a superposition principle. We will see in the sequel how to take advantage of this.

Example 2.1 Let us take $d = 1$ and $\Omega = (0; 1)$. The first equation (i) of (2.8) reads then

$$-\mu u'' = f$$

in $(0, 1)$. In the very particular case where f is a constant f_0, it is easy to obtain a closed-form solution:

$$u(x) = \underbrace{-\frac{f_0}{2\mu}x(x-1)}_{u_0(x)} + \underbrace{g_0(1-x) + g_1 x}_{u_g(x)},$$

where $g_0 := g(0)$ and $g_1 := g(1)$. In this case, we observe that a smooth (polynomial) solution exists and seems to be unique. This solution has been obtained by a superposition principle, which makes things easier: the coefficients of the polynomial are obtained with very few algebraic operations. A graphical representation of some solutions is depicted Fig. 2.5.

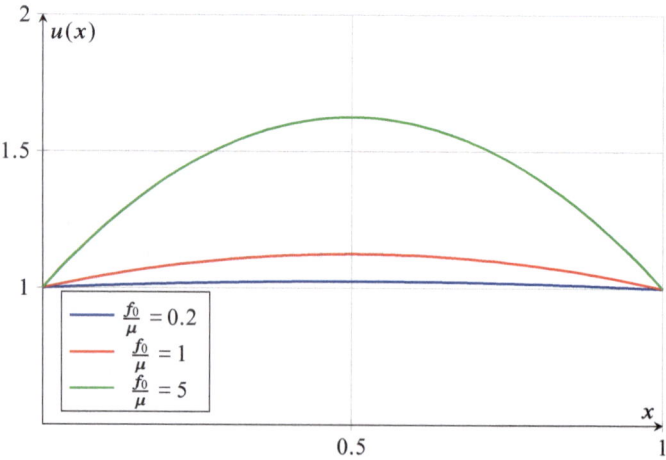

Fig. 2.5 Closed-form solution to Poisson's problem with Dirichlet boundary condition in one dimension, for different values of $\frac{f_0}{\mu}$, and $g = 1$

Now, suppose that the solution u to Problem (2.8) is in $\mathscr{C}^2(\overline{\Omega})$ and take $v \in \mathscr{C}^1(\overline{\Omega})$ a test function. We deduce from Eq. (2.8)–(i) and the Green formula (2.6):

$$\int_\Omega fv = -\mu \int_\Omega (\Delta u)v = \mu \int_\Omega \nabla u \cdot \nabla v - \mu \int_\Gamma (\partial_n u)v.$$

We now impose an additional condition on v, and we choose it such that it vanishes on the boundary Γ:

$$v|_\Gamma = 0.$$

So the boundary term $\int_\Gamma (\partial_n u)v$ vanishes. Set now

$$a(v, w) := \mu \int_\Omega \nabla v \cdot \nabla w, \quad L(w) := \int_\Omega fw,$$

for $v, w \in \mathscr{C}^1(\overline{\Omega})$. We just have shown that if $u \in \mathscr{C}^2(\overline{\Omega})$ is solution to Problem (2.8), then it solves also the following variational equation:

Find $u \in \mathscr{C}^2(\overline{\Omega})$ that satisfies $u|_\Gamma = g$ on Γ and

$$a(u, v) = L(v) \quad \text{for all } v \in \mathscr{C}^1(\overline{\Omega}) \text{ with } v|_\Gamma = 0. \tag{2.11}$$

First, notice here that the functional setting above is oldfashioned and looks very much like mathematics of the nineteenth century. So our first work will consist in providing a setting from the mathematics of the twentieth century, at least. This will allow particularly (1) to define a well-posed problem and guarantee it is well-posed and (2) to prove rigorously that the strong and weak problems are equivalent, thanks to the distribution theory.

2.3.2 Towards the True Weak Form

In fact, it appears clearly that the bilinear form and the linear form in the weak formulation (2.11) have meaning without the very strong continuity and differentiability assumptions previously stated. Let us extend their definition and introduce for this purpose $L^2(\Omega)$ as the Lebesgue space of real-valued square-integrable functions defined on Ω:

$$L^2(\Omega) := \left\{ v : \Omega \to \mathbb{R} \,\middle|\, \int_\Omega |v(x)|^2 \, dx < +\infty \right\}$$

endowed with the inner product

$$(v, w)_\Omega := \int_\Omega v(x)w(x)\,dx$$

for $v, w \in L^2(\Omega)$, and associated norm

$$\|v\|_{0,\Omega} := \left(\int_\Omega |v(x)|^2\,dx \right)^{\frac{1}{2}}.$$

With the above inner product, $L^2(\Omega)$ is a separable Hilbert space. For the proof, one can refer, for instance, to the corresponding chapter of H. Brezis about L^p spaces: [60, Theorem IV.8] (Fischer-Riesz Theorem) and [60, Theorem IV.13]. Now note that, with the assumption $f \in L^2(\Omega)$, the expression $L(w)$ is well-defined for any arbitrary $w \in L^2(\Omega)$. In fact, with the above notation, we have $L(w) = (f, w)_\Omega$.

Then define the simplest Sobolev space on Ω as

$$H^1(\Omega) := \left\{ v \in L^2(\Omega) \,\middle|\, D^\alpha v \in L^2(\Omega), |\alpha| = 1 \right\},$$

where $D^\alpha v$ is the distributional derivative of v to the multi-index α. We introduce on $H^1(\Omega)$ the following inner product:

$$(v, w)_{1,\Omega} := \sum_{|\alpha| \leq 1} (D^\alpha v, D^\alpha w)_\Omega$$

for $v, w \in H^1(\Omega)$ with the corresponding norm

$$\|v\|_{1,\Omega} := \left(\sum_{|\alpha| \leq 1} \|D^\alpha v\|_{0,\Omega}^2 \right)^{\frac{1}{2}}$$

for $v \in H^1(\Omega)$. This is still a separable Hilbert space. This is a consequence of the Hilbertian structure of $L^2(\Omega)$ and of the construction of the distributional derivative [232, Theorem 2.1] (see also [222]). Then just notice that $a(v, w)$ is well-defined for $v, w \in H^1(\Omega)$, which is a class of functions much larger than $\mathscr{C}^1(\overline{\Omega})$ (in two or three dimensions, it contains even functions that are not continuous, see e.g. [222]). To end the story, there remains to give a meaning to the value of functions on the boundary Γ.

2.3.3 An Insight into the Trace Theory and the True Weak Form

Indeed, it is not obvious to define the restriction (trace) on the boundary Γ for a function in $H^1(\Omega)$, since it may not even be continuous. For this purpose, we need the following result:

Theorem 2.3 (Trace) *Let Ω be a Lipschitz domain in $\mathbb{R}^d$. Denote $\Gamma := \partial\Omega$, its boundary. The trace mapping $\Upsilon : v \mapsto v|_\Gamma$, well-defined in $\mathscr{C}^0(\overline{\Omega})$, can be uniquely extended by density to the linear-bounded operator*

$$\Upsilon : H^1(\Omega) \to L^2(\Gamma).$$

Particularly, there holds the following trace inequality*:*

$$\|\Upsilon v\|_{0,\Gamma} \le C\|v\|_{1,\Omega} \tag{2.12}$$

for any $v \in H^1(\Omega)$, where $C > 0$ does not depend on v.

Proof See, for instance, [232, Theorem 4.2] for a careful construction of this trace mapping.   □

Thus, for a Lipschitz domain, functions in $H^1(\Omega)$ have their trace on the boundary well-defined in $L^2(\Gamma)$ and this allows to reformulate Problem (2.11) using a different setting, more friendly for analysis purposes. First, we set $V := H^1(\Omega)$ and

$$V_0 := H_0^1(\Omega) := \{v \in H^1(\Omega) \mid \Upsilon v = 0\},$$

the space of functions with vanishing trace on the boundary (the kernel of the trace mapping).

Now time has come to be accurate with the assumption on the boundary function g. With the assumption $u \in H^1(\Omega)$, the condition $u = g$ on Γ means that g should be the trace of a function in $H^1(\Omega)$. We define factually

$$H^{\frac{1}{2}}(\Gamma) := \{w \in L^2(\Gamma) \mid \exists v \in H^1(\Omega), w = \Upsilon v\},$$

as the range of the trace operator Υ. This is mandatory, since it can be shown that the image of $H^1(\Gamma)$ by Υ is a proper space of $L^2(\Gamma)$ [156, 157, 198, 232]. The notation $1/2$ in exponent is not casual and comes from the fact that this space can be characterized, for instance, using the Fourier transform and the quantity $1/2$ can be interpreted in terms of regularity [198, 232, 241] (broadly speaking these are functions more regular than $L^2(\Gamma)$ and less regular than $H^1(\Gamma)$). The space $H^{\frac{1}{2}}(\Gamma)$ is endowed with the induced norm

$$\|w\|_{\frac{1}{2},\Gamma} := \inf\{\|v\|_{1,\Omega} \mid v \in H^1(\Omega), \Upsilon v = w\}.$$

for $w \in H^{\frac{1}{2}}(\Gamma)$ [86, 232].

Let us now suppose that the boundary datum g satisfies

$$g \in H^{\frac{1}{2}}(\Gamma)$$

so that it is compatible with the regularity assumption expected for the solution u on the boundary. The weak form that corresponds to Poisson's problem with essential boundary condition (2.8) is finally

Find $u \in V$ that satisfies $\Upsilon u = g$ on Γ and

$$a(u, v) = L(v) \quad \text{for all } v \in V_0. \tag{2.13}$$

Now that we have a better setting involving weaker regularity assumptions on the solution and test functions, with vector spaces endowed with friendly properties (Hilbert spaces), we are ready to establish the well-posedness of the problem. This first step consists in writing an equivalent formulation with homogeneous boundary conditions.

2.3.4 Reformulation Using the Lifting Operator

We now reformulate Problem (2.13) and need first a *lifting* of the boundary datum $g \in H^{\frac{1}{2}}(\Gamma)$ i.e. a function $u_g \in H^1(\Omega)$ that coincides with g on the boundary: $\Upsilon u_g = g$. Previously, we introduced Problem (2.10), and this can be a way of getting explicitly this lifting, provided this problem admits a solution (we have not proven this yet): the solution u_g to Problem (2.10), if it exists, is called a harmonic lifting. We have already seen a harmonic lifting in one dimension (see Example 2.1):

$$u_g(x) = g_0(1 - x) + g_1 x,$$

which is a polynomial of degree 1 that belongs to the kernel of the Laplace operator in one dimension. Of course, this is not the only possibility to build a lifting of the boundary datum and, for instance,

$$u_g^1(x) = g_0(1 - x^2) + g_1 x^2, \qquad u_g^2(x) = g_0(1 - x) + g_1 x + 4\sin(\pi x),$$

also are candidate for a lifting: they both belong to $H^1(0, 1)$ and are equal to g for $x = 0, 1$. In fact, there exists an infinity of liftings, since any smooth enough function that coincides with g on the boundary Γ is a lifting. So uniqueness cannot be expected. This is not a problem since uniqueness will not be necessary in the following construction, but the issue of nonuniqueness needs to be treated with care.

Here, we present a more general and stronger result that ensures the existence of such a lifting. The main, delicate, point is to ensure that the lifting will be bounded by the boundary datum in the appropriate norms. We can do this in an elegant and expeditious form thanks to the Open Mapping Theorem of S. Banach that we recall below (for the proof, see the book of H. Brezis [60, Theorem 2.6], or see also [125, Lemma A.36]):

Theorem 2.4 (Open Mapping) *Let E and F be two Banach spaces, and let T be a linear continuous mapping from E to F. Suppose that T is surjective. Then there exists a constant $c > 0$ such that*

$$B_F(0, c) \subset T(B_E(0, 1)),$$

where $B_F(0, c)$ (resp. $B_E(0, 1)$) denotes the open ball in F of center 0 and of radius c (resp. the open ball in E of center 0 and of radius 1), and $T(B_E(0, 1))$ is the image of $B_E(0, 1)$ after application of T.

Now the following result will ensure the existence of a lifting for a datum $g \in H^{\frac{1}{2}}(\Gamma)$, with the required boundedness property.

Theorem 2.5 (Lifting) *Let Ω be a Lipschitz domain in $\mathbb{R}^d$ of boundary $\Gamma := \partial\Omega$. There exists $C > 0$ such that for all $w \in H^{\frac{1}{2}}(\Gamma)$, there is (at least) one function v_w in $H^1(\Omega)$ that verifies*

$$\Upsilon v_w = w, \quad \|v_w\|_{1,\Omega} \leq C\|w\|_{\frac{1}{2},\Gamma}. \tag{2.14}$$

Such a function v_w is called a lifting (or an extension) of w into $H^1(\Omega)$.

Proof We apply Theorem 2.4 to the Banach spaces

$$E = H^1(\Omega), \quad F = H^{\frac{1}{2}}(\Gamma),$$

and with $T = \Upsilon$ that is the trace operator. The trace operator Υ is linear, continuous and surjective, thanks to the Trace Theorem 2.3 and to the definition of $H^{\frac{1}{2}}(\Gamma)$ with its induced norm. So there exists $c_\Upsilon > 0$ such that

$$B_{H^{\frac{1}{2}}(\Gamma)}(0, c_\Upsilon) \subset \Upsilon(B_{H^1(\Omega)}(0, 1)).$$

Take now $w \in H^{\frac{1}{2}}(\Gamma)$ and define

$$\tilde{w} := \frac{c_\Upsilon w}{2\|w\|_{\frac{1}{2},\Gamma}} \in B_{H^{\frac{1}{2}}(\Gamma)}(0, c_\Upsilon) \subset \Upsilon(B_{H^1(\Omega)}(0, 1)).$$

From the above relationship, we deduce that there exists $\tilde{v} \in B_{H^1(\Omega)}(0, 1)$ such that $\Upsilon\tilde{v} = \tilde{w}$. Now we define

$$v_w := \frac{2\|w\|_{\frac{1}{2},\Gamma}}{c_{\Upsilon}} \tilde{v} \in H^1(\Omega).$$

It satisfies first $\Upsilon v_w = w$ and moreover

$$\|v_w\|_{1,\Omega} = \frac{2\|w\|_{\frac{1}{2},\Gamma}}{c_{\Upsilon}} \|\tilde{v}\|_{1,\Omega} \leq \frac{2}{c_{\Upsilon}}\|w\|_{\frac{1}{2},\Gamma}$$

so that the boundedness property in (2.14) is also satisfied. This ends the proof.   □

Remark 2.4 In fact, we can alternatively do the analysis of Poisson's problem with non-homogeneous boundary conditions without this above result and using only the definition of the norm $\|\cdot\|_{\frac{1}{2},\Gamma}$, see, for instance, [232, Chapter 4].

Remark 2.5 Another possibility to prove the above result is to make use of a constructive argument and to build explicitly a lifting which Sobolev norm is bounded by the norm of the datum, using the Fourier transform. This is done, for instance, in the book of W. McLean [198] (see also [86]).

So let $v_g \in H^1(\Omega)$ be a lifting of $g \in H^{\frac{1}{2}}(\Gamma)$ granted by the above Lifting Theorem 2.5. Suppose that $u \in V$ solves Problem (2.13). Define

$$u_0 := u - v_g \in V_0.$$

Since u is solution to Problem (2.13), there holds, for $v \in V_0$,

$$a(u_0, v) = a(u, v) - a(v_g, v) = L(v) - a(v_g, v).$$

Conversely, if u_0 solves

$$a(u_0, v) = L(v) - a(v_g, v),$$

then $u_0 + v_g(= u)$ solves Problem (2.13). Therefore, Problem (2.13) is equivalent to the following problem that involves only homogeneous Dirichlet boundary conditions:

Find $u_0 \in V_0$ such that

$$a(u_0, v) = L(v) - a(v_g, v) \quad \text{for all } v \in V_0. \tag{2.15}$$

2.3.5 Well-Posedness

To prove the well-posedness of Problem (2.15), the fundamental tool is the following Poincaré-Friedrichs inequality:

Theorem 2.6 *Let Ω be a bounded domain. Then there exists $c_P > 0$ such that*

$$\|v\|_{0,\Omega} \le c_P \|\nabla v\|_{0,\Omega}, \qquad \forall v \in H_0^1(\Omega). \tag{2.16}$$

Moreover, the constant c_P depends only on the diameter of Ω. As a result, the H^1-semi-norm is a norm on $H_0^1(\Omega)$, equivalent to the H^1-norm. Particularly, there exists $\alpha > 0$ such that:

$$\alpha \|v\|_{1,\Omega}^2 \le \|\nabla v\|_{0,\Omega}^2, \qquad \forall v \in H_0^1(\Omega), \tag{2.17}$$

where the constant $\alpha > 0$ depends only of the Poincaré-Friedrichs constant c_P.

Proof See for instance [232, Theorem 2.3] for the Poincaré-Friedrichs inequality (2.16) (see also [9, 60, 125, 222]). For the second assertion of the theorem, we notice that, for $v \in H_0^1(\Omega)$, there holds

$$\|v\|_{1,\Omega}^2 = \|v\|_{0,\Omega}^2 + \|\nabla v\|_{0,\Omega}^2 \le (c_P^2 + 1)\|\nabla v\|_{0,\Omega}^2,$$

which proves (2.17).   □

Then we prove in detail the following theorem:

Theorem 2.7 *Poisson's Problem (2.13) admits one unique solution. Moreover, there holds the stability bound:*

$$c\|u\|_{1,\Omega} \le \frac{1}{\mu}\|f\|_{0,\Omega} + \|g\|_{\frac{1}{2},\Gamma}, \tag{2.18}$$

where $c > 0$ depends only on the diameter of the domain Ω and is independent of μ.

Proof First, note that the space V_0 is a subspace of V. The Trace Theorem 2.3 ensures the trace operator Υ is continuous, so V_0 is closed. As a result, V_0 is still a Hilbert space. We check that the linear form $L(\cdot) - a(v_g, \cdot)$ is continuous on V_0.

Due to (2.17) in Theorem 2.6, the bilinear form $a(\cdot, \cdot)$ satisfies a coercivity (ellipticity) condition on V_0

$$a(v, v) \ge \mu\alpha\|v\|_{1,\Omega}^2, \qquad \forall v \in V_0, \tag{2.19}$$

where the ellipticity constant $\alpha > 0$ depends only on the Poincaré-Friedrichs constant c_P. So the bilinear form $a(\cdot, \cdot)$ is definite and positive on V_0. The bilinear form $a(\cdot, \cdot)$ is also symmetric and continuous on V_0. As a result, $a(\cdot, \cdot)$ is an inner product on V_0. Thus, thanks to the Riesz-Fréchet Representation Theorem (Theorem 2.2), there is a unique vector u_0 in V_0 such that

$$a(u_0, \cdot) = L(\cdot) - a(v_g, \cdot).$$

This means that Problem (2.15) admits one unique solution u_0, and, therefore, the function $u = u_0 + v_g$ is a solution to Problem (2.13).

Next, we need to establish the uniqueness of solutions to Problem (2.13) (there is no uniqueness for the lifting v_g): we suppose that $u_1 \in V$ and $u_2 \in V$ are two solutions to the Problem (2.13). Therefore, $\delta u := u_1 - u_2 \in V_0$ is solution to the variational formulation:

$$a(\delta u, v) = 0 \quad \text{for all } v \in V_0.$$

Once again, Riesz-Fréchet Representation Theorem permits to conclude that there is a unique solution $\delta u \in V_0$ to the above problem. Since $0 (\in V_0)$ is also solution, then necessarily, $0 = \delta u = u_1 - u_2$, which establishes uniqueness.

The *a priori* stability bound (2.18) is obtained as follows: let $u \in V$ be the unique solution to Problem (2.13) and set $u_0 := u - v_g \in V_0$, solution to Problem (2.15). We set $v = u_0 \in V_0$ in (2.15). We use the coercivity of $a(\cdot, \cdot)$, see (2.19), and apply Cauchy-Schwarz inequality:

$$\mu\alpha\|u_0\|_{1,\Omega}^2 \leq a(u_0, u_0) = L(u_0) - a(v_g, u_0) \leq \|f\|_{0,\Omega}\|u_0\|_{0,\Omega} + \mu\|v_g\|_{1,\Omega}\|u_0\|_{1,\Omega}.$$

We simplify the above expression:

$$\mu\alpha\|u_0\|_{1,\Omega} \leq \|f\|_{0,\Omega} + \mu\|v_g\|_{1,\Omega}.$$

We combine this last inequality to a triangular inequality and to the estimation on the lifting (2.14):

$$\|u\|_{1,\Omega} \leq \|u_0\|_{1,\Omega} + \|v_g\|_{1,\Omega}$$
$$\leq \frac{1}{\mu\alpha}\|f\|_{0,\Omega} + \left(\frac{1}{\alpha} + 1\right)\|v_g\|_{1,\Omega}$$
$$\leq \frac{1}{\mu\alpha}\|f\|_{0,\Omega} + C\left(\frac{1}{\alpha} + 1\right)\|g\|_{\frac{1}{2},\Gamma},$$

which is (2.18). $\square$

Remark 2.6 Remark that the above result ensures well-posedness in the Hadamard sense: thanks to the Riesz-Fréchet Representation Theorem, existence and uniqueness of the solution u are guaranteed. The stability bound (2.18) ensures also that the solution u is continuous with respect to the data (f, g), since our model problem is linear.

2.3.6 Back to the Strong Form and Strong-Weak Equivalence

Now let $u \in V$ be the solution to Problem (2.13). Let us show that u satisfies (2.8) in some sense, more general than the classical sense. First, the condition $\Upsilon u = g$ implies that (2.8)–(ii) holds almost everywhere on Γ. Then we take $\varphi \in \mathscr{D}(\Omega)$ and use the definition of the distributional divergence:

$$\langle \mathrm{div}\,(\nabla u), \varphi \rangle = -\langle \nabla u, \nabla \varphi \rangle.$$

But, since $\nabla u \in L^2(\Omega)$, ∇u is a regular distribution. Therefore, we can write, using this time (2.13)–(i)

$$\langle \nabla u, \nabla \varphi \rangle = \int_\Omega \nabla u \cdot \nabla \varphi = \int_\Omega f \varphi$$

to conclude that

$$-\langle \mathrm{div}\,(\nabla u), \varphi \rangle = \int_\Omega f \varphi$$

which means that $-\Delta u = f$ in $\mathscr{D}'(\Omega)$. Since f is a regular distribution, Δu is also a regular distribution, and using the Variational Lemma (1.1), this means that Eq. (2.8)–(i) holds almost everywhere in Ω. Conversely, we show in the same manner that if (2.8) is satisfied in the distributional sense, then u solves (2.13).

2.3.7 Equivalence with a Minimization Problem

Last but not least, following for instance [232, Lemma 4.3], we may view Problem (2.13) as a constrained minimization problem:

Proposition 2.3 *The unique solution to Problem (2.13) is also the unique minimizer on $H^1(\Omega)$ of the quadratic convex functional*

$$\mathcal{J}: H^1(\Omega) \ni v \mapsto \frac{1}{2}a(v, v) - L(v) \in \mathbb{R}$$

under the equality constraint $v|_\Gamma = g$.

Proof Let $u \in V$ be the unique solution to Problem (2.13) and let $v \in V$ that satisfies $v|_\Gamma = g$. We compute

$$\mathcal{J}(v) = \mathcal{J}(u+(v-u)) = \frac{1}{2}a(u, u) + \frac{1}{2}a(v-u, v-u) + a(u, v-u) - L(u) - L(v-u).$$

Since $v - u \in V_0$, and u solves (2.13), we deduce that

$$a(u, v - u) = L(v - u)$$

and thus

$$\mathcal{J}(v) = \mathcal{J}(u) + \frac{1}{2}a(v - u, v - u).$$

Moreover, because of the coercivity (ellipticity) condition (2.19) on V_0 ($v - u \in V_0$) we have the bound:

$$\mathcal{J}(v) \geq \mathcal{J}(u) + \frac{\mu\alpha}{2}\|v - u\|_{1,\Omega}^2,$$

which proves that u is the unique global minimizer of $\mathcal{J}$ in the affine subspace $\{v \in V \mid v|_\Gamma = g\}$. □

2.4 Poisson's Problem with a Neumann Boundary Condition

Let us see now how to model another boundary condition for Poisson's problem, where a flux is imposed on the boundary instead of the value of the solution itself. This flux depends on the gradient of the solution. Still Ω is an open subset of $\mathbb{R}^d$ ($d = 1, 2, 3$), supposed Lipschitz and bounded. To simplify the discussion, we suppose that Ω is a connected set. We still consider a physical parameter $\mu \in \mathbb{R}$ (the diffusion coefficient), with $\mu > 0$ and a source term $f : \Omega \to \mathbb{R}$, which is a prescribed scalar field in the domain.

2.4.1 From the Strong Form to the Weak Form

We introduce a boundary term $h : \Gamma \to \mathbb{R}$, which stands for a prescribed flux (a prescribed heat flux, or a prescribed boundary force for instance). Poisson's problem with a (pure) Neumann boundary condition reads

Find $u : \Omega \to \mathbb{R}$ solution to

$$\begin{cases} -\mu\Delta u = f & \text{in } \Omega, \quad (i) \\ \mu\partial_n u = h & \text{on } \Gamma. \quad (ii) \end{cases}$$

$$(2.20)$$

Example 2.2 Let us take $d = 1$ and $\Omega = (0; 1)$. The first equation (i) of (2.8) reads then

$$-\mu u'' = f$$

in $(0, 1)$. Since $n = (-1)$ at $x = 0$ and $n = (1)$ at $x = 1$, there holds

$$\partial_n u(0) = \nabla u \cdot n(0) = -u'(0), \qquad \partial_n u(1) = \nabla u \cdot n(1) = u'(1).$$

As a result, the second equation (ii) of (2.20) reads

$$-\mu u'(0) = h_0, \qquad \mu u'(1) = h_1,$$

with the convention $h_0 = h(0)$ and $h_1 = h(1)$. In the very particular case where f is a constant f_0, we can once again compute an explicit expression of the solution. First, from $u'' = -f_0/\mu$ we integrate and get

$$u'(x) = -\frac{f_0}{\mu}x + c_1$$

where $c_1 \in \mathbb{R}$ is a constant to be determined. This is the general form of the derivative, without the application of the boundary conditions.

The application of the two boundary conditions yield

$$-\mu c_1 = h_0, \qquad -f_0 + \mu c_1 = h_1.$$

Remark here, that, conversely to what happens for Dirichlet boundary conditions, we obtain an overdetermined system. It has a solution only if the following compatibility condition

$$-f_0 = h_0 + h_1$$

holds. This compatibility condition involves all the data (source term f on the bulk and boundary term h). We will get back to this point later on. Suppose that it is satisfied, then

$$c_1 = -\frac{h_0}{\mu} = \frac{h_1 + f_0}{\mu},$$

and

$$u'(x) = -\frac{f_0}{\mu}x - \frac{h_0}{\mu}.$$

We integrate to obtain the final solution and get

$$u(x) = -\frac{f_0}{2\mu}x^2 - \frac{h_0}{\mu}x + c.$$

Above the constant $c \in \mathbb{R}$ is arbitrary and we recover that, provided the compatibility condition is satisfied, the Neumann problem admits an infinity of solutions that form an affine space of dimension 1. Any particular solution can be recovered by setting a particular value of the constant, and in this case, u may satisfy a given extra property. For instance, with $c = 0$, there holds $u(0) = 0$. Here it is worth to note that the superposition principle is not straightforward to apply because of the compatibility condition. Let us see now what happens if the compatibility condition is not satisfied, and we can take, for instance, $\mu = 1$, $f_0 = 0$ and $h_0 = 0$, while we put the value $h_1 = 1$. Then the above solution reads

$$u(x) = c.$$

This means that $u'(x) = 0$ and then $u'(0) = u'(1) = 0$. So we cannot satisfy one of the Neumann boundary conditions. In fact, if this compatibility condition is not satisfied, there is an impossibility to satisfy all the euqations of (2.20).

Let us assume that u is regular enough, as previously for the Dirichlet boundary condition, for instance we take $u \in \mathscr{C}^2(\overline{\Omega})$, and we take v as a $\mathscr{C}^1(\overline{\Omega})$ test function. We start from the first equation (i) in (2.20) and apply the Green formula (2.6):

$$\int_\Omega \mu \nabla u \cdot \nabla v - \int_\Gamma \mu(\partial_n u)v = \int_\Omega fv.$$

We apply the Neumann condition $\mu \partial_n u = h$ on the boundary Γ:

$$\int_\Omega \mu \nabla u \cdot \nabla v - \int_\Gamma hv = \int_\Omega fv.$$

The second integral term has no more unknown u and can go to the right-hand side:

$$\int_\Omega \mu \nabla u \cdot \nabla v = \int_\Omega fv + \int_\Gamma hv. \tag{2.21}$$

As we did in the previous section for Dirichlet boundary conditions, we can provide an appropriate functional setting for the above weak formulation. We remark once again that the above bilinear form is well-defined provided that $u, v \in H^1(\Omega)(= V)$ and we can suppose the data $f \in L^2(\Omega)$ and $h \in L^2(\Gamma)$ to have the two terms at the right-hand side well-defined. Indeed, thanks to the Trace Theorem 2.3, there holds, for $v \in H^1(\Omega)$, $\Upsilon v \in L^2(\Omega)$ and so

$$\int_\Gamma hv = \int_\Gamma h(\Upsilon v) \in \mathbb{R}.$$

We use the same notations as previously for the bilinear form: $a(v, w) = \int_\Omega \nabla v \cdot \nabla w$, for $v, w \in V$. For the linear form associated with the source term f in the domain and the source term h on the boundary, we note it

$$L_n(v) = L(v) + \int_\Gamma hv = \int_\Omega fv + \int_\Gamma hv,$$

for $v \in V$. We are then ready to introduce the weak form associated with the Neumann problem:

Find $u \in V$ that satisfies

$$a(u, v) = L_n(v) \quad \text{for all } v \in V. \tag{2.22}$$

Because $\mathscr{C}^2(\overline{\Omega}) \subset \mathscr{C}^1(\overline{\Omega}) \subset V$, if $u \in \mathscr{C}^2(\overline{\Omega})$ is a strong solution to (2.20), then it verifies also the weak form (2.22). Let us now show, as in the first example, how a compatibality condition on the data appears, first as a necessary condition for the existence of a solution

Proposition 2.4 (Compatibility Condition) *Let $u \in V$ be a solution to the weak Neumann Problem (2.20), then there holds the compatibility condition*

$$L_n(1_\Omega) = \int_\Omega f + \int_\Gamma h = 0, \tag{2.23}$$

where 1_Ω denotes the constant function equal to 1 in Ω.

Remark 2.7 We can interpret the above Proposition as follows: if f and h do not satisfy the compatibility condition (2.23), the Neumann Problem (2.22) cannot have any solution.

Proof Let us take $v = 1_\Omega \in V$ in the weak form (2.22), then $\nabla v = 0$ and we get

$$0 = a(u, 1_\Omega) = L_n(1_\Omega) = \int_\Omega f + \int_\Gamma h.$$

$\square$

Remark 2.8 The compatibility condition (2.23) can be recovered directly from the strong form (2.20) after the application of Gauss Theorem (Proposition 2.1).

2.4.2 The Deny-Lions Theorem

Before establishing well-posedness, we need an important result to prove an ellipticity property. Indeed, the previous Poincaré-Friedrichs inequality (Theorem 2.6) cannot be used because it relies on the property that functions in $H_0^1(\Omega)$ vanish on the boundary. So we state here a more general and powerful result that generalizes Theorem 2.6. For the proof of this result, see, for instance, [232, Theorem 7.1].

Theorem 2.8 (Deny-Lions) *Let* $\Omega \subset \mathbb{R}^d$ *be an open, bounded, connected, Lipschitz domain. Let* $j : H^1(\Omega) \to \mathbb{R}$ *be a linear-bounded functional such that* $j(1_\Omega) \neq 0$. *Then there exists* $C > 0$ *such that*

$$\|v\|_{0,\Omega} \leq C \left(\|\nabla v\|_{0,\Omega}^2 + |j(v)|^2 \right)^{\frac{1}{2}} \tag{2.24}$$

for all $v \in H^1(\Omega)$.

Remark first that we recover Theorem 2.6 as a straightforward consequence of the above result: it suffices to take $j(v) := \int_\Gamma v$. However, Theorem 2.6 needs fewer assumptions in fact (only an open and bounded domain) and has a direct constructive proof (the proof of Theorem 2.8 is based on a contradiction argument).

Let us emphasize also on the importance of all the assumptions on the domain Ω for the validity of Theorem 2.8. Particularly, the domain needs to be bounded and Lipschitz in order to apply the Rellich-Kondrachov Theorem (compacity of the injection of $H^1(\Omega)$ into $L^2(\Omega)$) that is at the core of the proof. In fact, the Lipschitz assumption can be weakened and every domain that has the H^1-extension property is still valid. The result can be adapted for a domain that is not connected, but in this case we need multiple functionals such as j (see [232] for details).

Remark 2.9 Another way to recover the above result is to derive it as an application of the Peetre-Tartar lemma; see [125].

2.4.3 Well-Posedness

As in [51], we introduce the linear form

$$m : V \ni v \mapsto m(v) := \frac{1}{|\Omega|} \int_\Omega v \in \mathbb{R}$$

that computes the mean value of a function v in the domain Ω. We check that m is well-defined and continuous, since

$$|m(v)| \leq \frac{1}{|\Omega|} \int_\Omega 1|v| \leq \|v\|_{0,\Omega} \leq \|v\|_{1,\Omega},$$

where we used the Cauchy-Schwarz inequality. We introduce as well the projection operator

$$P : V \ni v \mapsto P(v) := v - m(v)1_\Omega \in V_m,$$

where 1_Ω still denotes the constant function equal to 1 on Ω and where

$$V_m := \{v \in V \mid m(v) = 0\},$$

is the kernel of m. We check that

$$m(P(v)) = m(v) - m(m(v)1_\Omega) = m(v) - m(v)m(1_\Omega) = 0,$$

since $m(1_\Omega) = 1$. As a result, P projects the functions in V into the subspace of functions with zero mean that is in direct sum to the vector line spanned by constant functions.

Let us prove first the following Lemma:

Lemma 2.1 *Let us suppose that the compatibility condition* (2.23) *holds* ($L_n(1_\Omega) = 1$). *Then the Problem* (2.22) *is equivalent to the following problem:*

Find $u \in V$ that satisfies

$$a(P(u), v) = L_n(v) \quad \text{for all } v \in V_m. \tag{2.25}$$

Proof Let us take $u \in V$ solution to (2.22) and take $v \in V_m \subset V$. Then there holds

$$a(u, v) = L_n(v).$$

Then we decompose $u = P(u) + m(u)1_\Omega$ and use $a(m(u)1_\Omega, v) = 0$ ($\nabla(1_\Omega) = 0$) so that

$$a(P(u), v) = L_n(v), \quad \forall v \in V_m.$$

Conversely, let us suppose now that $u \in V$ solves the above variational equation. Then, for all $v \in V$, we decompose it as $v = P(v) + m(v)1_\Omega$. We use again the property that $a(\cdot, \cdot)$ vanishes for constant functions and the compatibility condition $L_n(1_\Omega) = 0$ to write

$$a(u, v) = a(P(u), P(v)) = L_n(P(v)) = L_n(P(v)) + m(v)L_n(1_\Omega) = L_n(v).$$

So $u \in V$ is solution to (2.25), and we have proven the equivalence between the two variational equations, provided the compability condition is satisfied.   □

Then, we need the following Poincaré inequality:

Lemma 2.2 (Poincaré) *For Ω an open, bounded, connected, Lipschitz domain, there exists $C > 0$ such that*

$$\|P(v)\|_{0,\Omega} \le C\|\nabla v\|_{0,\Omega}, \qquad \forall v \in V. \tag{2.26}$$

Proof We apply the Deny-Lions Theorem 2.8, see [232, Corollary 7.3]. It suffices to take $j = m$ and we verify that $j(1_\Omega) = 1$. Then there exists $C > 0$ such that, for $v \in V$

$$\|v\|_{0,\Omega} \le C\left(\|\nabla v\|_{0,\Omega}^2 + |m(v)|^2\right)^{\frac{1}{2}}.$$

We apply the above inequality for $P(v) \in V$, use $\nabla P(v) = \nabla v$ as well as $m(P(v)) = 0$, and get (2.26). $\qquad\qquad\square$

Then we can state our main result of well-posedness, which is a consequence of the previous statements (see also [232, Proposition 7.7]).

Theorem 2.9 *If the compatibility condition (2.23) is not satisfied, Problem (2.22) does not admit any solution. Conversely, if (2.23) holds, the set of solution to Problem (2.22) is*

$$\{u_m + C1_\Omega \mid C \in \mathbb{R}\}$$

with $u_m \in V_m$ the unique solution to Problem (2.25). Moreover, every solution $u \in V$ to Problem (2.22) verifies the bound:

$$c\left\|u - \frac{1}{|\Omega|}\int_\Omega u\right\|_{1,\Omega} \le \frac{1}{\mu}\left(\|f\|_{0,\Omega} + \|h\|_{0,\Gamma}\right), \tag{2.27}$$

with $c > 0$.

Proof From Proposition 2.4, we have seen that if (2.23) is not satisfied, there cannot be any solution. Now suppose that the compatibility condition (2.23) holds. Then from Lemma 2.1, Problems (2.22) and (2.25) are equivalent.

Since m is continuous, V_m is a Hilbert space, as a closed subspace of V. Moreover, since Ω is a bounded Lipschitz domain, the Poincaré inequality of Lemma 2.2 ensures that $a(\cdot, \cdot)$ defines an inner product on V_m. Then, we can apply the Riesz-Fréchet Representation Theorem 2.2: Problem (2.25) admits one solution u_m, unique in V_m, which is also a solution to Problem (2.22) (remark that $P(u_m) = u_m$).

Suppose now that $u \in V$ is a solution to Problem (2.22). Then $u - u_m$ satisfies

$$a(P(u - u_m), v) = a(P(u), v) - a(u_m, v) = L_n(v) - L_n(v) = 0,$$

for all $v \in V_m$. The above variational problem has a unique solution in V_m, which is 0. This means that $0 = P(u - u_m)$. As a result

$$P(u) = P(u_m) = u_m,$$

and thus

$$u = P(u) + m(u)1_\Omega = u_m + m(u)1_\Omega.$$

This proves that the set of solutions is indeed

$$\{u_m + C1_\Omega \mid C \in \mathbb{R}\}.$$

Let us establish finally the a priori bound. Let $u \in V$ be a solution to Problem (2.22) and let us remember first that $P(u) = u_m$ to bound

$$c\mu \|P(u)\|_{1,\Omega}^2 = c\mu \|u_m\|_{1,\Omega}^2 \leq a(u_m, u_m) = L_n(u_m)$$

where we used (2.26). Then we apply Cauchy-Schwarz inequality and the trace inequality to bound

$$L_n(u_m) \leq \left(\|f\|_{0,\Omega} + \|h\|_{0,\Gamma}\right)\|u_m\|_{1,\Omega}.$$

We combine the two above bounds and get

$$c\mu \|P(u)\|_{1,\Omega} \leq \left(\|f\|_{0,\Omega} + \|h\|_{0,\Gamma}\right),$$

which is (2.27). This ends the proof. □

Remark 2.10 Remark that the compatibility condition is needed in the above proof for the equivalence between Problems (2.22) and (2.25), but that Problem (2.25) remains well-posed even if the compatibiity condition is not satisfied.

2.4.4 Back to the Strong Form

Let now $u \in V$ be a solution to the weak form of Neumann problem (Problem (2.22)). Let us first take $\varphi \in \mathscr{D}(\Omega)$ a test function. We proceed as in Sect. 2.3.6 and conclude first that $-\Delta u = f$ in $\mathscr{D}'(\Omega)$. Since f is a regular distribution, Δu is also a regular distribution, and using the Variational Lemma (1.1), this means that Eq. (2.20)–(i) holds almost everywhere in Ω. To recover the Neumann boundary condition almost everywhere, we need a stronger result which is described as follows:

First, we need to define the space of vector-valued square-integrable functions that admit a square-integrable distributional divergence

$$H(\text{div}; \Omega) := \left\{ \psi \in L^2(\Omega; \mathbb{R}^d) \,\middle|\, \text{div}\, \psi \in L^2(\Omega) \right\}$$

and in this case the divergence of ψ verifies the following relationship:

$$\int_\Omega (\text{div}\, \psi)\, \varphi = -\int_\Omega \psi \cdot \nabla\varphi, \quad \forall \varphi \in \mathscr{D}(\Omega).$$

The space $H(\text{div}; \Omega)$ is endowed with the norm

$$\|\psi\|_{H(\text{div};\Omega)} := \left(\|\psi\|_{0,\Omega}^2 + \|\text{div}\, \psi\|_{0,\Omega}^2 \right)^{\frac{1}{2}}.$$

Then we need to introduce $H^{-\frac{1}{2}}(\Gamma)$ that we define as the topological dual of $H^{\frac{1}{2}}(\Gamma)$. We denote by $\langle \cdot, \cdot \rangle_\Gamma$ the duality pairing in $H^{-\frac{1}{2}}(\Gamma) \times H^{\frac{1}{2}}(\Gamma)$ and denote by $\|\cdot\|_{-\frac{1}{2},\Gamma}$ the dual (operator) norm defined as

$$\|\varphi\|_{-\frac{1}{2},\Gamma} := \sup_{w \in H^{\frac{1}{2}}(\Gamma)} \frac{\langle \varphi, w \rangle_\Gamma}{\|w\|_{\frac{1}{2},\Gamma}},$$

for $\varphi \in H^{-\frac{1}{2}}(\Gamma)$.

The following intermediate result is a consequence of the Gauss Theorem 2.1 and of the lifting theorem (Theorem 2.5) (see [150, Theorem 2.5] or [86, Proposition 2.17]):

Proposition 2.5 *Let Ω be a Lipschitz domain in $\mathbb{R}^d$, of boundary $\Gamma := \partial\Omega$, then the mapping $\Upsilon_n : \phi \mapsto \phi \cdot n$, defined for $\mathscr{D}(\overline{\Omega}; \mathbb{R}^d)$, can be extended uniquely into a linear-bounded operator*

$$\Upsilon_n : H(\text{div}\,; \Omega) \to H^{-\frac{1}{2}}(\Gamma).$$

The above Proposition 2.5 combined with a density argument allows to extend the previous Green Theorem (Proposition 2.2) to a more general setting, as follows (see [86, 150, 181]):

Proposition 2.6 *Let Ω be a Lipschitz domain in $\mathbb{R}^d$, of boundary $\Gamma := \partial\Omega$. For every function $\psi \in H^1(\Omega)$ and $\phi \in H(\text{div}; \Omega)$, there holds $(\phi \cdot n) \in H^{-\frac{1}{2}}(\Gamma)$ as well as*

$$-\int_\Omega \text{div}\, (\phi)\psi = \int_\Omega \phi \cdot \nabla\psi - \langle \phi \cdot n, \Upsilon\psi \rangle_\Gamma.$$

Now let us take once again $u \in V$ a solution to (2.22) and w a test function on the boundary Γ, with $w \in H^{\frac{1}{2}}(\Gamma)$. We denote $v_w \in H^1(\Omega)$ its lifting, as granted by Theorem 2.5. We remember the previous considerations, and particularly

$$\text{div } \nabla u(= \Delta u) = f \in L^2(\Omega),$$

so we have $\nabla u \in H(\text{div}; \Omega)$. We apply Proposition 2.6 with $\phi = \nabla u$, $\psi = v_w$ and get

$$0 = \int_\Omega \text{div }(\nabla u)v_w + \int_\Omega \nabla u \cdot \nabla v_w - \langle \nabla u \cdot n, \Upsilon v_w \rangle_\Gamma.$$

Since u solves (2.22) and since we have already established that $-\Delta u = f$ holds almost everywhere on Ω, there remains from the above identity

$$0 = \int_\Gamma hw - \langle \partial_n u, w \rangle_\Gamma,$$

where we used $\Upsilon v_w = w$. From this, we deduce that in fact $\partial_n u \in L^2(\Omega)$ and

$$\partial_n u = h \quad \text{a.e. on } \Gamma.$$

We conclude that $u \in V$, solution to (2.22) is also a solution to (2.27).

Remark 2.11 In fact, we realize here that with the above extension of the Green formula, the assumption $h \in L^2(\Gamma)$ we made at the beginning can be relaxed, and all the above analysis can be extended without difficulty if we suppose only $h \in H^{-\frac{1}{2}}(\Gamma)$, see [232].

2.4.5 Equivalence with a Minimization Problem

Last but not least, we can also view Problem (2.22) as a minimization problem:

Proposition 2.7 *If the compatibility condition (2.23) holds, then the quadratic functional*

$$\mathcal{J} : H^1(\Omega) \ni v \mapsto \frac{1}{2}a(v, v) - L_n(v) \in \mathbb{R}$$

admits a unique minimum and the set of its minimizers is exactly the set of the solutions to Problem (2.22). If the the compatibility condition (2.23) does not hold, the functional $\mathcal{J}$ admits no minimum on $H^1(\Omega)$.

Proof Suppose first that the compatiblity condition (2.23) holds. Let $u \in V$ be a solution to Problem (2.22) and let $v \in V$. We compute

$$\mathcal{J}(v) = \mathcal{J}(u+(v-u)) = \frac{1}{2}a(u, u) + \frac{1}{2}a(v-u, v-u) + a(u, v-u) - L_n(u) - L_n(v-u)$$

and notice first that $a(u, v - u) = L_n(v - u)$ because of (2.22) $(v - u \in V)$. So

$$\mathcal{J}(v) = \mathcal{J}(u) + \frac{1}{2}a(v - u, v - u).$$

Moreover, because of the Poincaré inequality (2.26) there holds

$$\mathcal{J}(v) \geq \mathcal{J}(u) + \mu\alpha\|P(v - u)\|_{1,\Omega}^2,$$

which proves that the set of solutions to (2.22) and of minimizers to $\mathcal{J}$ coincide. Suppose now that the compatibility condition (2.23) is not satisfied: set $\beta = L_n(1_\Omega) \neq 0$ and then $\varphi_n = (1/\beta)n1_\Omega \in H^1(\Omega)$, for $n \geq 0$. Since

$$\mathcal{J}(\varphi_n) = -\frac{1}{\beta}nL_n(1_\Omega) = -n,$$

we have that $J(\varphi_n) \to -\infty$ when $n \to +\infty$, which proves that $\mathcal{J}$ has no minimum on $H^1(\Omega)$. $\square$

Remark 2.12 We recover here something similar to what is known in finite dimension, more precisely for quadratic functionals in finite-dimensional vector spaces [10].

2.5   An Insight into Regularity

Here we provide minimal information about the expected regularity of the solution u to Poisson's Problem with Dirichlet boundary conditions (Problem (2.8)). Indeed, from the weak form (2.13), we know that u belongs to $H^1(\Omega)$ but it is the least information about the regularity of u. Indeed, a function in $H^1(\Omega)$ may not be very smooth and for many problems, u is expected to be rather smooth, see, for instance, the one dimensional Example 2.1, where the solution is $\mathscr{C}^\infty([0, 1])$.

Notably, the one dimensional case is friendly, because the regularity of u depends only of the regularity of the source term f [60]. In dimension two or three, the maximal regularity of the solution, for Poisson's problem with Dirichlet boundary condition, depends notably of the regularity of the boundary.

This section summarizes the main results in two dimensions, following a presentation in [58]. For more insight into regularity and singularities, we recommend first, for instance [9, 125, 157, 232, 240], at introductory level, and [94, 99, 127, 128, 156, 231] or even [27, 96–98, 172] for more advanced topics. First, we introduce new Sobolev spaces, useful to characterize the regularity of weak solutions to boundary value problems. Then, we provide some regularity results.

2.5.1 The Family of Sobolev Spaces of Index m

In fact, the definition of the Sobolev space $H^1(\Omega)$ we have seen in 2.3.2 is easy to generalize to take into account higher order derivatives, see e.g. [5, 59, 198]. Indeed, for every exponent $m \in \mathbb{N}$, the Sobolev space of order m on Ω is defined as

$$H^m(\Omega) := \left\{ v \in L^2(\Omega) \,\middle|\, D^\alpha v \in L^2(\Omega), |\alpha| \leq m \right\},$$

where $D^\alpha v$ is the distributional derivative of v to the multi-index α. We adopt the usual convention: $H^0(\Omega) := L^2(\Omega)$, and we recover obviously the definition seen in 2.3.2 for $H^1(\Omega)$. We introduce on $H^m(\Omega)$ the following inner product:

$$(v, w)_{m,\Omega} := \sum_{|\alpha| \leq m} (D^\alpha v, D^\alpha w)_\Omega$$

for $v, w \in H^m(\Omega)$ with the corresponding norm

$$\|v\|_{m,\Omega} := \left(\sum_{|\alpha| \leq m} \|D^\alpha v\|_{0,\Omega}^2 \right)^{\frac{1}{2}}$$

for $v \in H^m(\Omega)$.

Basically, if a function v belongs to $L^2(\Omega)$, it has a little bit of regularity. For $\Omega = (-1, 1)$, it can be for instance a Heaviside step function, but not a delta Dirac distribution (see previous Chap. 1). If in addition, some high-order derivatives of this function still belong to $L^2(\Omega)$, this function is even more regular. Notably, observe that

$$\cdots \subset H^m(\Omega) \subset \cdots \subset H^2(\Omega) \subset H^1(\Omega) \subset L^2(\Omega).$$

As a result, the Sobolev index m allows in some sense to charactarize the regularity of a function v.

Regularity in a more classical sense can then be obtained using, for instance, the Sobolev Embedding Theorems, see e.g. Theorem B.42 (Morrey) and Corollary B.43 in [125] (as well as the Figure B.1 in the aforementioned reference for the illustration of functions that belong, or not, to $H^1(\Omega)$ in one dimension).

2.5.2   Regularity and Singularities for Polygons

First, if Ω is a bounded convex polygon, the solution u to Problem (2.13) belongs to $H^2(\Omega)$, see [125, Theorem 3.12] and [94, 156, 157]. When Ω is a nonconvex polygon, singularities appear at reentrant corner. We summarize the situation below, following a presentation by S. Brenner [58].

To simplify, we assume that $\mu = 1$, that $g = 0$, so that we have a homogeneous Dirichlet condition on the boundary Γ, and that Ω is a bounded polygonal domain in $\mathbb{R}^2$ with at least one re-entrant angle. We recall that the source term f belongs to $L^2(\Omega)$. We denote by $\omega_1, \ldots, \omega_J$ internal angles of Ω that correspond to re-entrant corners and satisfy $\pi < \omega_j < 2\pi$, with $s_j \in \mathbb{R}^2$ the corresponding vertices. In this case, the unique solution $u \in H_0^1(\Omega)$ to Problem (2.13) (with $g = 0$) has the representation

$$u = \sum_{j=1}^{J} \kappa_j u_j^S + u^R. \tag{2.28}$$

Above, the regular part is $u^R \in H^2(\Omega) \cap H_0^1(\Omega)$. The singular functions $u_1^S, \ldots, u_J^S$ are defined as follows:

$$u_j^S(r_j, \theta_j) := \varphi_j(r_j) r_j^{\frac{\pi}{\omega_j}} \sin\left(\frac{\pi}{\omega_j}\theta_j\right). \tag{2.29}$$

Above, we use polar coordinates (r_j, θ_j) at the corresponding vertex s_j, such that ω_j is spanned by the two half lines $\theta_j = 0$ and $\theta_j = \omega_j$. The functions φ_j are smooth cutoff functions equal to one in a neighborhood of 0 and with a support small enough.

Example 2.3 For the L-shape domain depicted Fig. 2.6, the expression of the singular function at the reentrant angle $\omega(= \omega_1)$ is

$$u^S(r, \theta) = r^{\frac{2}{3}} \sin\left(\frac{2}{3}\theta\right), \tag{2.30}$$

since $\omega = 3\pi/2$ and where we omitted the indices and the cutoff function, to simplify. We represent it in Fig. 2.7.

The coefficients κ_j in the representation formula (2.28) can be obtained using the so-called extraction formula provided below:

$$\kappa_j := \frac{1}{\pi} \left(\int_\Omega f u_{-j}^S + \int_\Omega u \Delta u_{-j}^S \right) \tag{2.31}$$

Fig. 2.6 A typical L-shape domain Ω with the re-entrant angle ω

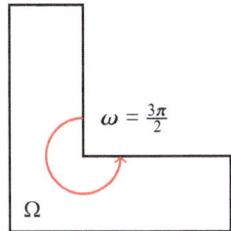

$\omega = \frac{3\pi}{2}$

Ω

Fig. 2.7 Sketch of the singular function u^S at the re-entrant corner of a L-shape domain

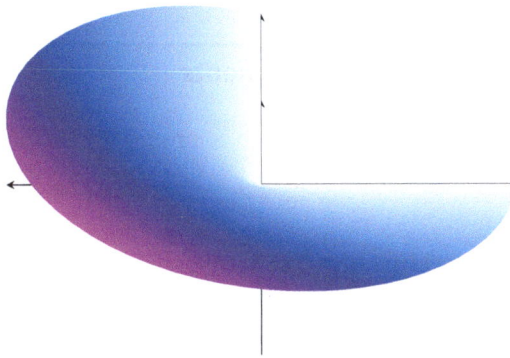

where the dual singular functions u^S_{-j} have the following expression:

$$u^S_{-j}(r_j, \theta_j) = \varphi_j(r_j) r_j^{-\frac{\pi}{\omega_j}} \sin\left(\frac{\pi}{\omega_j}\theta_j\right).   \tag{2.32}$$

For elasticity equations, the coefficients κ_j are called the stress intensity factors and they play an important role in linear fracture mechanics.

Finally, let us mention the regularity estimate below:

$$\|u^R\|_{2,\Omega} + \sum_{j=1}^{J} |\kappa_j| \leq C\|f\|_{0,\Omega},   \tag{2.33}$$

with $C > 0$. It can be proven that singular functions u^S_j cannot belong to $H^2(\Omega)$: see particularly [9] for a simple explanation and the references mentioned above and in [58] for more details. This prevents the global solution u to belong to $H^2(\Omega)$.

Remark 2.13 For a domain with a crack (angle equal to 2π at the crack tip) and small strain elasticity in two dimensions, see, for instance, [74, 157, 184] and references therein for useful information.

2.5.3 Why Regularity Is Important for Numerical Solving

For reasons that will appear clearly at the end of Chap. 4, standard finite element methods are somehow unefficient in case of singularities, particularly methods based on polynomials of order two or higher. In some simple situations, the precise knowledge of the singularities can be directly incorporated to modify the finite element method (as in [58], and see also [12,13,24,55,100,101,106,122]). However, a standard way to handle such singularities is now adaptive mesh refinement, based on a posteriori error estimators. This will be the objective of Chap. 7.

2.6 Further Comments

This section presents some complements and extensions on the topic of boundary value problems.

2.6.1 Lipschitz Domains

More details about Lipschitz domains can be found in [198, 232] particularly, but also in classical texts such as [156, 203]. Non-Lipschitz domains with cups are studied in e.g. [119]. Functional spaces and boundary value problems on fractal domains are studied in e.g. [2–4, 208].

2.6.2 Sobolev Spaces

Recent references, at introductory level, to enter smoothly into the topic of Sobolev spaces are, among others, [9, 107, 198, 232]. Classical textbooks are for instance [5, 60, 156, 197, 241]. For a historical perspective, see e.g. [201, 241].

2.6.3 Partial Differential Equations

Poisson's problem described in this section is an example of scalar linear elliptic partial differential equation. A detailed treatment of general elliptic partial differential equations and their approximations is generally presented in classical finite element textbooks, for instance [59, 87, 125, 222]. Other well-known scalar elliptic partial differential equations are the reaction-diffusion equation and the advection-diffusion equation.

Many more partial differential equations exist, and the majority of them require much more sophisiticated tools to study their well-posedness. Recent references on the topic are, for instance [127–129, 186, 202, 219, 232, 235], and a classical one is the series of books by R. Dautray and J.L. Lions; see [102] for the first volume.

At the introductive level, see particularly [9] for a detailed description of important examples of partial differential equations, with derivation from basic physical principles and a mathematical study of their properties. For recent perspectives about modelling (digital twinning) see, for instance, [166].

2.6.4 Boundary Conditions

Dirichlet and Neumann boundary conditions are the most emblematic ones in the study of boundary value problems. For further references about their mathematical study, see [172–175]. Nevertheless, in fact, there exists much more conditions on the boundary, mostly motivated to take into account the underlying physics. The most common one are mixed boundary conditions where a Dirichlet boundary condition is imposed on a portion of the boundary, and a Neumann condition is imposed on another portion, and Robin (or Fourier or impedence) boundary conditions; see e.g. [125, 232] for a detailed description of both. For Ventcel boundary conditions, that generalize Robin boundary conditions, see, for instance, [53]. For an example where boundary conditions lead to an ill-posed problem (the Cauchy problem), see [42].

Another class of boundary conditions are non-linear and come from contact (Signorini condition) and friction laws (Tresca or Coulomb friction). An example of such a condition (Signorini condition) is provided in Chap. 6. See also e.g. [86, 120, 121, 152, 164, 181, 248].

2.6.5 Interface Conditions and Domain Decomposition

In numerical simulation, another situation where conditions on a part of the boundary play a major role are interface problems (see the subsection Problems below for an example with details).

This can be motivated by multiphysics applications where two different partial differential equations occupy two different domains, and their unknowns communicate at the interface between the two domains (the terminology heterogeneous domain decomposition is also common [49]). See, for instance, [134, 135] for fluid-structure interaction.

Interface conditions can also appear artificially in domain decomposition for parallel computing, when the physical domain is splitted somehow arbitrarily to distribute the numerical solving procedure over all the nodes of a parallel architecture. In this case, the boundary conditions that appear for each subdomain may have no physical meaning, but are needed for communication between the subdomains within an iterative procedure. For classical references on the topic, the reader can refer to [132, 144, 221]. For optimized methods based on Robin boundary conditions, see [143] and for time-dependent problems, see [145].

Problems

This section describes an interface problem and is inspired from [9]. The solution is provided at the end of this book. A second part for this problem follows in Chap. 5.

Let Ω be an open bounded Lipschitz domain in $\mathbb{R}^2$, with boundary $\Gamma := \partial\Omega$ and outward unit normal n to the boundary. We consider first the datum

$$k : \Omega \to \mathbb{R}$$

and a second datum

$$f : \Omega \to \mathbb{R}.$$

We want to solve the following elliptic boundary value problem:

$$\begin{cases} -\text{div}\,(k\nabla u) = f \text{ in } \Omega, \\ u = 0 \qquad\qquad \text{on } \Gamma, \end{cases} \tag{2.34}$$

where the unknown is $u : \Omega \to \mathbb{R}$.

The second datum f is the source term and we suppose that $f \in L^2(\Omega)$. The first datum k represents a conductivity, which can have different values inside of Ω. This allows to model an heterogeneous medium. We suppose that k is a measurable function, and that there are two positive real numbers, denoted by k_m et k_M, that verify

$$0 < k_m \leq k(x) \leq k_M$$

for almost every x inside Ω.

2.1 Which problem do we recover if we suppose that $k(x) = 1$ for every $x \in \Omega$?

2.2 Provide the weak form associated with Problem (2.34).

2.3 Prove that Problem (2.34) admits one unique solution.

Now let us suppose there exists a partition of Ω in two subdomains Ω_1 and Ω_2 ($\overline{\Omega} = \overline{\Omega_1} \cup \overline{\Omega_2}$ and $\Omega_1 \cap \Omega_2 = \emptyset$), as in Fig. 2.8. We suppose that both subdomains are open connected sets. We denote by $\Sigma := \partial\Omega_1 \cap \partial\Omega_2$ the interface between the two subdomains and we suppose it smooth. We denote by n_i the outward unit normal on Σ, exterior to Ω_i ($i = 1, 2$). We denote by $\Gamma_i := \Gamma \cap \partial\Omega_i$ the external boundary of Ω_i, and thus $\partial\Omega_i = \Gamma_i \cup \Sigma$ ($i = 1, 2$).

We denote by k_i the restriction of k on each subdomain Ω_i, $i = 1, 2$:

$$k_i(x) = k(x) \quad \text{for } x \in \Omega_i.$$

Fig. 2.8 Interface problem.
Partition of the domain Ω into
two subdomains

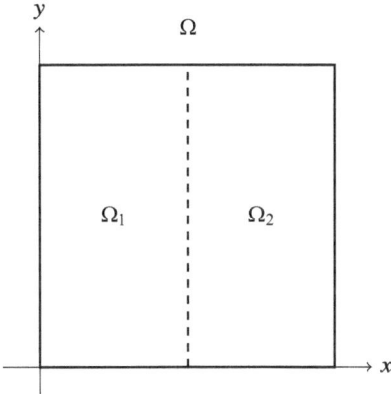

Since k is not assumed to be continuous, there can be a discontinuity of the
conductivity k at the interface Σ. Let us introduce the two problems, for $i = 1, 2$:
Find $u_i : \Omega \rightarrow \mathbb{R}$ solution to

$$\begin{cases} -\text{div}\, (k_i \nabla u_i) = f \text{ in } \Omega, \\ u_i = 0 \qquad\qquad \text{ on } \Gamma_i, \end{cases} \tag{2.35}$$

with the interface conditions

$$\begin{cases} u_1 = u_2 & \text{on } \Sigma, \\ k_1 \nabla u_1 \cdot n_1 + k_2 \nabla u_2 \cdot n_2 = 0 \text{ on } \Sigma. \end{cases} \tag{2.36}$$

2.4 Show that, from Problem (2.35), for every v a regular enough test function in
Ω, u_1 and u_2 solve

$$\sum_{i=1}^{2} \int_{\Omega_i} k_i \nabla u_i \cdot \nabla v - \int_{\Sigma} (k_1 \nabla u_1 \cdot n_1 + k_2 \nabla u_2 \cdot n_2) v = \sum_{i=1}^{2} \int_{\Omega_i} f v.$$

2.5 Use (2.36) to get a weak formulation satisfied by u_1 and u_2.

Low-Order Lagrange Finite Elements

3

Prior to discretizing the Poisson problem, we need to introduce fundamental concepts of the finite element method, with one of the simplest finite elements: the first-order (piecewise linear) Lagrange finite element on a simplex.

3.1 Outline

Let us get back to the pipeline presented Fig. 2.1 for numerical simulation. Previous chapter detailled the process of mathematical modelling. It seems to be a fastiduous process and in fact, it is, above all if one is faced with this activity for the first time and wants to understand it in depth. But in fact, in real life, many engineers have hacks to almost bypass it. And, if one gets more experience in mathematics, one gets aware of the different points that may cause difficulty when one is faced with a new model. But let us suppose now that modelling is done and let us try to use a computer to approximate its solution. So we get to the second step of the pipeline, which is approximation. See Fig. 3.1.

We detail below the first ingredients and steps related to the numerical approximation with finite element technology.

3.1.1 Variational Approximation or (Ritz-)Galerkin Methods

Let us remember that the solution u to Poisson's problem with Dirichlet boundary conditions in weak form, Problem (2.13), belongs to a vector space V of admissible functions. The main idea of the finite element method is to take into account this variational structure (the weak form) into the discretization process. So the vector space V, of infinite dimension, is approximated by a space V^h of finite dimension. This is a characteristic of all (Ritz-)Galerkin methods, or variational discretization methods. As a result, an approximated solution u^h to Problem (2.13) is sought in V^h,

F. Chouly, *Finite Element Approximation of Boundary Value Problems*,
Compact Textbooks in Mathematics, https://doi.org/10.1007/978-3-031-72530-2_3

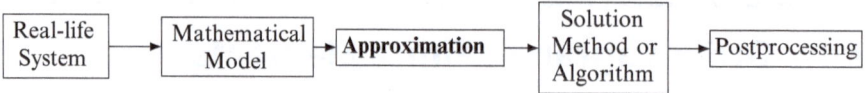

Fig. 3.1 The above pipeline depicts the global process behind a numerical simulation. We are now describing the process of approximation of the mathematical model

and the goal is to build V^h so that there is a chance that the solution u^h delivered by the computer be close enough to the (true) continuous solution u.

In this aspect, the finite element method differs from other discretization techniques, also of common use in mathematics, physics, engineering, such as finite differences, finite volumes, colocation methods, lattice Boltzmann, etc., that proceed directly from the strong form of the partial differential equation, and that do not care that much if there is a function space behind (at least for the implementation).

At this stage, the main point is the following: let $\varphi_1, \ldots, \varphi_N$ be a basis of V^h (which is a vector space of finite dimension), then every element v^h of V^h can be written

$$v^h = V^1 \varphi_1 + \ldots + V^N \varphi_N, \tag{3.1}$$

and reconstructed from a finite collection of real numbers $(V^1, \ldots, V^N)$. Here and now, the computer can be involved because he is capable of processing these numbers (up to roundoff errors).

3.1.2   Meshing the Domain and the Boundary

The first step for approximation with the finite element method is to build a mesh of the domain of interest Ω when the problem has to be solved. Many finite element solvers have this capability or can import a mesh produced by an external mesh generator. The idea behind meshing is to split the domain into small cells (or subdomains) of small size and (generally) simple shape. For instance, simplicial meshes are made of triangles in dimension two and tetrahedra in dimension three. Fig. 3.2 provides an example of a mesh in dimension two. The notation $\mathcal{T}^h$ can be used to design this mesh, and the letter T stands for a generic simplex in the mesh. Then, the index h represents generally the mesh size, which means the diameter of the largest simplex in the mesh: the smaller is h, the larger is the number of cells.

3.1.3   Piecewise Polynomial Functions on the Mesh

Once the mesh $\mathcal{T}^h$ is generated, the discrete space V^h can be designed by setting the value of a function v^h at all the nodes of the mesh (the vertices of the triangles in two dimensions). Then, to recover the value on the simplices T, we can simply use

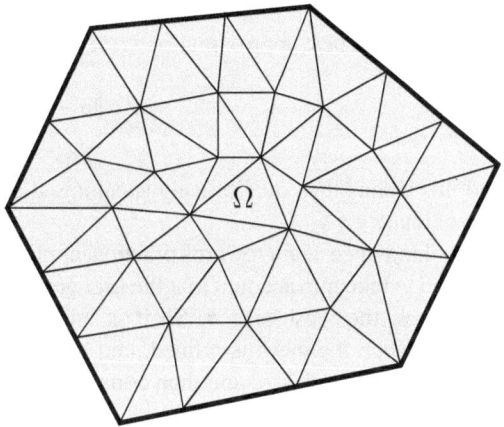

Fig. 3.2 Example of a valid triangular mesh of a polygonal domain Ω

linear interpolation. For this purpose, we introduce on each simplex T the space of polynomials of order 1:

$$\mathbb{P}_1(T) := \text{Span}\,(1, x_1, \ldots, x_d) = \{\alpha_0 + \alpha_1 x_1 + \cdots + \alpha_d x_d \mid (\alpha_0, \ldots, \alpha_d) \in \mathbb{R}^{d+1}\}.$$

This is a vector space of dimension $d + 1$. And then define the discrete space as

$$V^h := \left\{ v^h \in \mathscr{C}^0(\overline{\Omega}) \,\middle|\, v^h|_T \in \mathbb{P}_1(T), \forall\, T \in \mathcal{T}^h \right\}. \tag{3.2}$$

This is the vector space of the lowest order Lagrange finite elements. In this special case, the decomposition (3.1) for v^h in V^h can be rewritten

$$v^h = v^h(a_1)\varphi_1 + \ldots + v^h(a_N)\varphi_N, \tag{3.3}$$

where $a_1, \ldots, a_N$ are the nodes of the mesh (the vertices of the simplices). So the coefficients to determine, or degrees of freedom, correspond directly to the value of the function v^h at each node. The basis functions $\varphi_1, \ldots, \varphi_N$ are hat functions, equal to one for the corresponding node, and to zero at the other mesh nodes.

3.1.4 Approximation Error

Since the space V^h defined previously is used to approximate the solution $u \in V$, and since this is a proper space of V, the approximate solution u^h will differ generally from u. For instance, going back to Example 2.1 in Chap. 2: if the exact solution u is a polynomial of order equal to two, no piecewise linear function u^h in V^h will match perfectly with it.

So there is a discrete error, or approximation error, that corresponds to the distance between u and u^h. It is usual, and natural, to use the Hilbert space structure of V and the norm $\|\cdot\|_{1,\Omega}$ to measure this error. A lower bound for the discrete error is given by the best approximation error

$$\inf_{v^h \in V^h} \|u - v^h\|_{1,\Omega}$$

and this means that, qualitatively speaking, the space V^h should be rich enough to approximate u.

Looking once again towards practical applications, and for numerical simulation, we need to take into account that the user generally controls, up to his computational ressources, the mesh size h. So it is interesting to know if the discrete error is reduced when the mesh is refined, and how much. This is all the most useful than increasing the mesh size and then computing the solution using a finer mesh can be time-consuming.

One early mathematical breakthrough in the finite element community has been results of polynomial approximation in Sobolev spaces (see e.g. [59, 87, 115, 125, 220] and references therein). For the piecewise linear Lagrange finite element, the local estimates can be of the form:

$$\|\nabla u - \nabla(\mathscr{I}^T u)\|_{0,T} \leq Ch_T \tag{3.4}$$

for u regular enough. In the above equation, where h_T is the size of the simplex T, $\mathscr{I}^T$ is the Lagrange interpolation of u on T, and $C > 0$ is the constant that depends on u and the "shape" of T (C can be large for elongated simplices). In the next chapter, this will allow to prove that finite element methods converge in $O(h)$.

3.2 Simplicial Meshes

We introduce here a common way to subdivise the domain into cells of simple shape that will be used later on to approximate functions that are solutions of partial differential equations.

3.2.1 Simplices

We recall that the convex hull generated by some points is the smallest convex set that contains these points. Let us first recall the definition of a simplex.

Definition 3.1 Let $d \geq 1$ and $0 \leq \delta \leq d$. A δ-simplex T of $\mathbb{R}^d$ is the convex hull generated by $\delta+1$ points $(a_i)_{i=0,\dots,\delta}$ of $\mathbb{R}^d$. The δ-simplex T is non-degenerate if the set of vectors $(a_i - a_0)_{i=1,\dots,\delta}$ are linearly independent, otherwise T is degenerate.

We call the points $(a_i)_{i=0,\dots,\delta}$ the nodes or the vertices of T. Note that δ-simplices are closed subsets. They verify the following property:

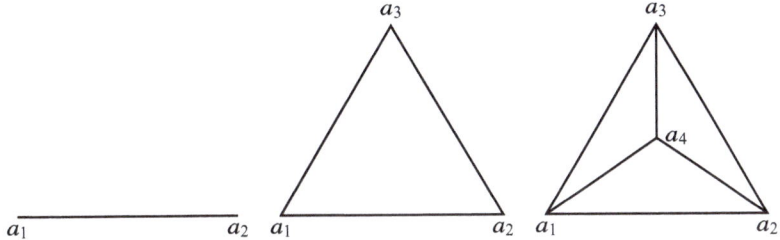

Fig. 3.3 Examples of 1-simplex (left), 2-simplex (middle), and 3-simplex (right)

Proposition 3.1 *The boundary of a δ-simplex is constituted of δ + 1 (δ − 1)-simplices.*

In fact, for non-degenerate simplices, we recover familiar notions:

1. A 0-simplex is a point. We can also call it a node or a vertex.
2. A 1-simplex is a straight segment between two points. We can also call it an edge.
3. A 2-simplex is a triangle.
4. A 3-simplex is a tetrahedron.

Figure 3.3 provides an illustration in various dimensions.
Let T be a δ-simplex. We define:

1. $\mathcal{N}_T$ the set of its $\delta + 1$ vertices $(a_i)_{i=0,\dots,\delta}$.
2. If $\delta \geq 2$, $\mathcal{E}_T$ the set of its edges, which are 1-simplices.
3. If $\delta \geq 3$, $\mathcal{A}_T$ the set of its (triangular) faces, which are 2-simplices.

Last but not least, for a simplex T, we denote by $|T|$ its measure. We also denote h_T, its diameter, that can be defined formally as

$$h_T := \max_{x,y \in T} \|x - y\|$$

where $\| \cdot \|$ denotes the Euclidean norm in $\mathbb{R}^d$. As well, we denote by ρ_T the radius of the largest ball contained in the simplex T and a_ρ the centre of this ball.

3.2.2   Definition of a Simplicial Mesh

Let us suppose that $\Omega \subset \mathbb{R}^d$ is a polytopal domain (polygonal domain in dimension two or a polyhedron in dimension three). We start by providing a precise definition of what is a simplicial mesh:

Definition 3.2 A simplicial mesh $\mathcal{T}^h$ of the domain Ω is a collection of d-simplices that are subsets of $\overline{\Omega}$. We call $T \in \mathcal{T}^h$ a generic simplex in the mesh (segment for $d = 1$, triangle for $d = 2$, and tetrahedron for $d = 3$). We denote by $\mathcal{A}^h$ the corresponding collection of (triangular) faces (which is an empty set when $d < 3$), by $\mathcal{E}^h$ the corresponding collection of edges (which is an empty set when $d = 1$), and by $\mathcal{N}^h$ the corresponding collection of nodes. Moreover, this collection of simplices should satisfy two properties:

1. The mesh covers exactly the domain:

$$\overline{\Omega} = \bigcup_{T \in \mathcal{T}^h} T.$$

2. For every pair of distinct simplices (T, T') in $\mathcal{T}^h \times \mathcal{T}^h$, their intersection should be either empty or verify

$$T \cap T' \in \mathcal{N}^h \cup \mathcal{E}^h \cup \mathcal{A}^h.$$

Let us comment a little bit on this definition. The first thing is that each simplex T covers a portion of the closure of the domain Ω ($T \subset \overline{\Omega}$) and $\mathcal{T}^h$ is simply a set that collects all the simplices:

$$\mathcal{T}^h = \{T_1, \ldots, T_M\},$$

where

$$M := \text{card}\,(\mathcal{T}^h) \tag{3.5}$$

is the number of simplicial cells in the mesh. The tradition within the Finite Element community is to denote a generic cell in $\mathcal{T}^h$ by T without any index notation. The set of nodes can be defined formally as

$$\mathcal{N}^h := \bigcup_{T \in \mathcal{T}^h} \mathcal{N}_T.$$

In the same spirit, we define

$$\mathcal{E}^h := \bigcup_{T \in \mathcal{T}^h} \mathcal{E}_T$$

and

$$\mathcal{A}^h := \bigcup_{T \in \mathcal{T}^h} \mathcal{A}_T.$$

We define precisely the size h of the mesh $\mathcal{T}^h$ as follows:

$$h := \max_{T \in \mathcal{T}^h} h_T,$$

where h_T is the diameter of the simplex T, as introduced previously. See finally Fig. 3.2 for an example of valid simplicial mesh.

3.2.3 Families of Simplicial Meshes and Shape Regularity

Another tradition in the finite element community is to consider a family of meshes. For uniform refinement, it is usual to index them using the mesh size and to consider, for instance, a family $(\mathcal{T}^h)_{h>0}$ of simplicial meshes of the domain Ω. We say that a family of meshes is regular in Ciarlet's sense [87, 88] if there exists a real $\sigma > 0$ such that

$$h_T \leq \sigma \rho_T, \quad \forall T \in \mathcal{T}^h. \tag{3.6}$$

This means, roughly speaking, that the shape of the cells is uniformly controlled within a mesh and for all the meshes of the family. This assumption prevents simplices to degenerate when the mesh is refined.

Remark 3.1 For other equivalent criteria of shape regularity in the context of simplicial meshes, see, for instance, [57].

3.3 First-Order Lagrange Finite Elements

First, we introduce:

$$\mathbb{P}_1(T) := \text{Span} \, (1, x_1, \ldots, x_d) = \{\alpha_0 + \alpha_1 x_1 + \cdots + \alpha_d x_d \mid (\alpha_0, \ldots, \alpha_d) \in \mathbb{R}^{d+1}\}$$

on each simplex T: this is the space of multi-variate polynomials of order 1, a vector space of dimension $d + 1$.

The Lagrange finite element space of degree (or order) 1 is (see e.g. [59, 87, 127, 220]):

$$V^h := \left\{ v^h \in \mathscr{C}^0(\overline{\Omega}) \, \middle| \, v^h|_T \in \mathbb{P}_1(T), \forall T \in \mathcal{T}^h \right\}. \tag{3.7}$$

It means that each function in V^h is a continuous, piecewise polynomial function of order 1, and its restriction to each simplex is a polynomial of order 1. Therefore, $(V^h)_{h>0}$ is a family of finite-dimensional vector spaces indexed by h, built from the above family $(\mathcal{T}^h)_{h>0}$ of simplicial meshes of the domain Ω.

The space V^h defined above is H^1-conformal in the following sense:

Proposition 3.2 *Let Ω be a polygonal domain in $\mathbb{R}^2$ or a polyhedral domain in $\mathbb{R}^3$ and let V^h be the Lagrange finite element space defined in (3.7). There holds*

$$V^h \subset H^1(\Omega).$$

Proof See Problem 3.1.   □

3.3.1 Basis of Shape Functions

We denote by $N \in \mathbb{N}$ the dimension of the vector space V^h, and by $\varphi_1, \ldots, \varphi_N$ the nodal basis of shape functions of V^h. The corresponding finite element nodes, or degrees of freedom, are denoted by $a_1, \ldots, a_N$, so that there holds the relationship

$$\varphi_i(a_j) = \delta_{ij} \tag{3.8}$$

where $1 \leq i, j \leq N$, and δ_{ij} is Kronecker's symbol ($\delta_{ij} = 1$ if $i = j$ and $\delta_{ij} = 0$ if $i \neq j$ for $1 \leq i, j \leq N$).

We may sometimes split the set of finite element nodes into two subsets: first $I := 1, \ldots, N_{int}$ for the internal nodes, which means that:

$$a_i \in \Omega, \quad \forall i \in I.$$

Then, we define $B := (N_{int} + 1), \ldots, N$ the complementary set of nodes that belong to the boundary Γ:

$$a_i \in \Gamma, \quad \forall i \in B.$$

Note we have chosen to index the nodes such that the first N_{int} nodes be internal nodes.

3.3.2 The Lagrange Interpolation Operator

The Lagrange interpolation operator, or Lagrange interpolator, associated with V^h will be denoted by $\mathscr{I}^h$. It is defined as follows, for any continuous function v in $\overline{\Omega}$:

$$\mathscr{I}^h(v) := \sum_{i=1,\ldots,N} v(a_i)\varphi_i \in V^h.$$

We denote by $\mathscr{I}^h_\Gamma$ the Lagrange interpolation operator associated to the boundary Γ: it is defined for any $w \in \mathscr{C}(\overline{\Gamma})$ by

$$\mathscr{I}_\Gamma^h(w) := \sum_{i \in B} w(a_i)\, \Upsilon \varphi_i,$$

where Υ is the trace operator introduced in Chap. 2. We will need later on the following useful property (see [125]):

Proposition 3.3 *The trace operator and the Lagrange interpolator are commuting, in the following sense:*

$$\Upsilon(\mathscr{I}^h v) = \mathscr{I}_\Gamma^h(\Upsilon v), \quad \forall v \in \mathscr{C}(\overline{\Omega}) \cap H^1(\Omega). \tag{3.9}$$

Proof See Problem 3.2. □

3.3.3 Estimates for Lagrange Interpolation

Here we state a basic local error estimate for Lagrange interpolation. For this purpose, it will be convenient to use the following notation for Sobolev semi-norms in $H^m(T)$, $m \geq 0$, $T \in \mathcal{T}^h$:

$$|v|_{m,T} := \left(\sum_{|\alpha|=m} \|D^\alpha v\|_{0,\Omega}^2 \right)^{\frac{1}{2}}$$

for $v \in H^m(T)$.

Theorem 3.1 *Let $u \in H^2(\Omega)$ where $\Omega \subset \mathbb{R}^d$. The following interpolation estimate holds*

$$|u - \mathscr{I}^h u|_{m,T} \leq C h_T^{2-m} |u|_{2,T}, \tag{3.10}$$

for $m = 0, 1$ and every simplex $T \in \mathcal{T}^h$. The positive constant C is independent of T and of h_T, but depends of the shape regularity constant $\sigma > 0$ in (3.6).

Proof See [59, 87, 115, 125, 127, 220, 222], for instance. □

3.4 Further Comments

This section provides some extra references about meshing, finite element spaces, and numerical approximation methods.

3.4.1   Numerical Approximation Technologies

Among classical methods used from the twentieth century and the invention of the first computers, we can mention, at least, the fast Fourier transform (FFT), finite differences, finite elements, finite volumes, and spectral methods. A recent monograph of M. Gander and F. Kwok introduces most of them; see [146].

There is still a great amount of activity relative to (Ritz-)Galerkin methods alternative to the finite element method. Let us just mention a few of them (see e.g. [18] for a general framework):

- Discontinuous Galerkin (DG) methods, see, for instance, the pioneering works [19, 188] and the recent monograph [109]
- Recent polytopal methods for very general meshes (cells are polytopes), such as conforming polygonal finite elements, see e.g. [239]; hybrid discontinuous Galerkin (HDG), see e.g. [90]; hybrid high-order (HHO) methods, see e.g. [89, 90, 108, 187]; weak Galerkin method, see e.g. [112]; virtual element method (VEM), see e.g. [40, 187]; smooth finite element method (SFEM), see e.g. [191, 207]; and modified discontinuous Galerkin with static condensation, see e.g. [194]
- Discontinuous Petrov Galerkin (DPG) methods, see e.g. [104, 105]
- IsoGeometric analysis (IGA) and variants, see e.g. [22, 95, 205], motivated by the link between computer aided design and numerical simulation
- Unfitted finite elements or geometrically nonconforming finite elements, where the mesh boundary and the domain boundary do not match, as it occurs in fictitious domain methods, the eXtended finite element method (XFEM) or the cut finite element method (cutFEM), see e.g. [54, 66, 118, 153, 165, 199, 206, 217]
- Reduced basis techniques, such as reduced order modelling (ROM) or proper orthogonal decomposition (POD), see e.g. [155, 168, 180, 204]
- Spectral methods, see e.g. [44] or [71]
- Wavelet-based discretization, see e.g. [48, 50, 91, 92, 200]
- Ritz methods based on deep learning and neural networks, see e.g. [47, 192]

3.4.2   Meshing

A classical reference about meshing is the monograph of P. Frey and P.L. Georges; see [141]. In addition, some classical textbooks about finite element approximation detail this aspect, for instance, [125].

3.4.3 Zoology of Finite Elements

Many different finite elements have been invented since the early years of this method (approximately 1950). There is even a periodic table of the finite elements.[1] In fact, for Lagrange finite elements, any polynomial order $k \geq 1$ can be considered, and the detailed general construction of the finite element vector space in this case is detailed, for instance, in [9, 125, 222]. Other well-known finite elements are, for instance, the Crouzeix-Raviart, the Raviart-Thomas, and the Nédélec finite elements; see [125]. Special finite elements have been designed for viscous fluids (Stokes and Navier-Stokes equations) and are detailed, for instance, in [150]. For thin structures like shells, see [76]. More and more finite elements can be found in e.g. [9, 20, 59, 61, 87, 125, 127–129, 176, 238, 249].

3.4.4 Finite Element Librairies

If you have no previous experience with finite elements, the Matlab scripts described [8] in [146] allow to understand fully all the implementation aspects, with the less possible technicalities, and keeping close to the mathematical theory. Particularly, they allow to understand fully how to transform the discrete weak form into a linear system. Furthermore, a mesh generator in Matlab is presented in [216].

For more than two decades now, high-level and open-source finite element libraries have been released by leading groups in mathematics, computer science, and/or mechanics that allow us to make use of the language of weak forms directly and that allow to solve efficiently rather complex problems (three-dimensional, non-linear, multiphysics, with possibly many degrees of freedom). They can avoid entering into implementation aspects such as assembly, quadratures, transformation to the reference element, etc., such as described in detail in [125, 249]. We can mention, for instance (in alphabetical order), FEniCS [193], FreeFEM++ [167], GetFEM [224], MooAFEM [171], or scikit-fem [159]. Other large-scale modern academic and/or industrial codes are, for instance, Code Aster,[2] FEEL++,[3] LifeX,[4] or MEF++.[5]

Problems

3.1 Let Ω be a polygonal domain in $\mathbb{R}^2$ or a polyhedral domain in $\mathbb{R}^3$ and let V^h be the Lagrange finite element space defined in (3.7). Prove that there holds

---

[1] https://www-users.cse.umn.edu/~arnold/femtable/.

[2] See https://code-aster.org/.

[3] See https://github.com/feelpp/feelpp.

[4] See https://lifex.gitlab.io/.

[5] See https://giref.ulaval.ca/.

$$V^h \subset H^1(\Omega).$$

3.2 Prove that the trace operator and the Lagrange interpolator are commuting, in the following sense:

$$\Upsilon(\mathscr{I}^h v) = \mathscr{I}_\Gamma^h(\Upsilon v), \quad \forall v \in \mathscr{C}(\overline{\Omega}) \cap H^1(\Omega). \tag{3.11}$$

The Standard Finite Element Method

4

Now we combine the two previous chapters to present the standard method to discretize Poisson's problem with Lagrange finite elements. Particular emphasis is made on how to handle the non-homogeneous Dirichlet boundary condition, first, and then the pure Neumann boundary condition.

4.1 Outline

We are going from the second step of our pipeline to the third step; see Fig. 4.1. A central idea of the Finite Element method is to combine:

1. The weak formulation (2.13) of the mathematical model (Chap. 2)
   and
2. The finite element finite-dimensional vector space V^h (Chap. 3).

This allows to recover a solution u^h that belongs to V^h and that is expected to be reasonably close to the exact solution u, if V^h is reasonably designed. In fact, u^h is obtained from a column vector U that contains all the values of u^h at the nodes of the mesh. A linear system is built from the weak formulation combined with V^h (Galerkin procedure), with a large size matrix, called usually the stiffness matrix, and the vector U as an unknown.

The practice of finite elements consists, to say it in a nutshell, in writing this stiffness matrix in a computer program (assembly), and then to run an algorithm that solves a linear system. In fact, the computer can only deliver a vector $\tilde{U}$, in general (but not always), close to the "theoretical" vector U. So the solution displayed by a computer, let us call it $\tilde{u}^h$ differs from the "theoretical" u^h. The numerical error is the difference between $\tilde{u}^h$ and u^h in an appropriate discrete norm. In our case, this difference comes from the solver for the linear system, especially if an iterative solver is used, and from roundoff errors. For more complex problems and more

© The Author(s), under exclusive license to Springer Nature Switzerland AG 2025
F. Chouly, *Finite Element Approximation of Boundary Value Problems*,
Compact Textbooks in Mathematics, https://doi.org/10.1007/978-3-031-72530-2_4

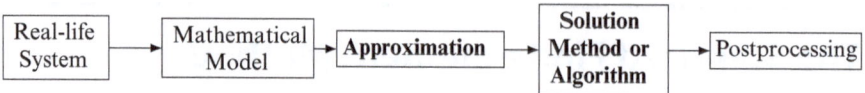

Fig. 4.1 The above pipeline depicts the global process behind a numerical simulation. We are ending with the second step of approximation and going to the third step

sophisticated numerical methods, other numerical errors coming from solvers for non-linear equations and from numerical integration can invite themselves.

4.1.1   Dirichlet Boundary Condition and Discrete Lifting

The standard method for imposing a non-homogeneous Dirichlet boundary condition relies on a discrete lifting or direct nodal imposition [31, 125, 128, 236].

For this purpose, we introduce a discrete approximatin of the Dirichlet boundary condition g, and we define $g^h := \mathscr{I}_\Gamma^h(g)$, where $\mathscr{I}_\Gamma^h$ denotes the Lagrange interpolant on the trace space of V^h. A discrete counterpart of Problem (2.13) is then:

$$\text{Find } u^h \in V^h \text{ that satisfies } u^h|_\Gamma = g^h \text{ and}$$

$$a(u^h, v^h) = L(v^h) \quad \text{for all } v^h \in V_0^h, \tag{4.1}$$

where $V_0^h := V^h \cap V_0$ is the subspace of V^h of discrete functions that vanish on the boundary Γ. As for the continuous case, Problem (4.1) is equivalent to find the unique minimizer on V^h of the quadratic convex functional $\mathcal{J}(\cdot)$ under the equality constraint $v^h|_\Gamma = g^h$.

Let $\mathcal{L}^h g^h \in V^h$ be a discrete lifting of g^h. It can be obtained, for instance, by setting $\mathcal{L}^h g^h(a_i) = g(a_i)$ if a_i is a boundary node ($a_i \in \Gamma$), and $\mathcal{L}^h g^h(a_i) = 0$ if a_i is an interior node ($a_i \in \Omega$). An equivalent formulation of Problem (4.1) is:

$$\text{Find } u^h \in V^h \text{ of the form } u^h = \mathcal{L}^h g^h + u_0^h, \text{ with } u_0^h \in V_0^h \text{ solution to}$$

$$a(u_0^h, v^h) = (f, v^h)_\Omega - a(\mathcal{L}^h g^h, v^h) \quad \text{for all } v^h \in V_0^h(\Omega). \tag{4.2}$$

Problem (4.2) is a discrete counterpart of Problem (2.15).

4.1.2   Mathematical Analysis of the Discrete Problem

Using the same argument as for the continuous problem, it can be established that the discrete Problem (4.1) is well-posed and an optimal H^1-error estimate can be derived. More precisely, let us suppose that the solution u to Problem (2.13) is regular enough. Let u^h be the solution to Problem (4.1), there holds:

$$\|u - u^h\|_{1,\Omega} \le Ch,$$

where the constant $C > 0$ does not depend of the mesh size h. This error estimate is in the natural (Sobolev) norm. It is qualified of *optimal* since the approximation error is in $O(h)$, as the interpolation error provided in Chap. 3: the error is of the same magnitude as if we would be able to interpolate directly the exact solution u with its value at the mesh nodes. In practice, it means that, if the global mesh size is reduced twice, then the error is reduced twice (linear convergence rate).

It is an important result: not only it proves that it is worth refining the mesh and that if the mesh is fine enough, the discrete error is reduced, but it also provides a speed of convergence, that may allow, for instance, a heuristics to stop the uniform refinement of the mesh.

4.1.3 Transforming a Boundary Value Problem into a Matrix Problem

Problem 4.2 can be recasted in matrix form as

$$K_I U_I = F - K_{BI} G, \qquad U_B = G.$$

where, broadly speaking, K. are blocks built from a matrix K associated to the differential operator (Laplace operator here), U is a column vector that collects all the nodal values of the solution, and F and G are column vectors that correspond to the source term in the bulk and to the Dirichlet condition on the boundary, respectively. This allows an implementation on a given computer, since computers are gifted to manipulate matrices and to solve linear systems.

The number of lines of the linear system is in $O(h^{-d})$, so uniform refinement of the mesh is counterbalanced by increasing the size of the global system. For a practitioner, this limits the minimal value of h that can be achieved, which will be determined at the end by the computer architecture, particularly its speed and its memory. We will see in Chap. 7 how this viewpoint can be improved.

4.1.4 Limitations of This Technology

The method of nodal imposition, or discrete lifting, though simple and efficient, is not easy to generalize to more complex boundary conditions (see Chap. 6 for contact conditions as an example). Furthermore, it is closely related to Lagrange finite elements on a conforming mesh and is not suited to other discretization approaches, such as those mentioned at the end of Chap. 3, in which degrees of freedom are not directly linked to the value of the discrete solution u^h at some nodes located on the boundary Γ. In the next chapter, we will present an alternative technique, among others, to overcome these issues.

4.2   The Standard FEM with a Discrete Lifting

In this section, we will study thoroughly the finite element discretization of Poisson's problem (2.8). We will describe the most standard procedure, that is presented, or at least mentioned, in all the basic textbooks and classnotes, and that is based upon a discrete lifting operator. To simplify the presentation, we suppose that, from now, the Lipschitz domain Ω is a polygon, for $d = 2$, or a polyedron (for $d = 3$), so as to avoid technical issues related to the approximation of curved boundaries. We will limit ourselves also to simplicial meshes and Lagrange finite element spaces, as described in the previous chapter (Chap. 3).

We use the finite element space V^h built from Lagrange elements of degree 1 defined in Chap. 3. We define also

$$V_0^h := V^h \cap V_0 = \{v^h \in V^h \mid v^h = 0 \text{ on } \Gamma\}$$

as the vector space of functions with vanishing discrete trace.

First, we describe the formulation of the standard method to enforce the non-homogeneous Dirichlet boundary condition (2.8)-(ii). Then we analyse it: we will prove that the discrete problem has a unique solution, and that the discrete solution converges towards the continuous one with an optimal rate. Finally, we describe its implementation. This presentation is inspired from the (first) monograph of A. Ern and J.-L. Guermond [125, Section 3.2, Section 8.4]. To simplify, we make the following assumptions:

- In Eq. (2.8)-(ii), the Dirichlet datum g is continuous:

$$g \in \mathscr{C}^0(\Gamma) \cap H^{\frac{1}{2}}(\Gamma).$$

- There exists a lifting of g in $H^1(\Omega)$ that is continuous (Theorem 2.5 does not guarantee that this property holds necessarily). As before, we note v_g this lifting:

$$v_g \in \mathscr{C}^0(\overline{\Omega}) \cap H^1(\Omega), \quad \Upsilon v_g = g.$$

Since g is continuous, we can take its Lagrange interpolant:

$$g^h := \mathscr{I}_\Gamma^h(g).$$

A discrete counterpart of Problem (2.13) is then:

$$\text{Find } u^h \in V^h \text{ that satisfies } \Upsilon u^h = g^h \text{ on } \Gamma \text{ and}$$

$$a(u^h, v^h) = L(v^h) \quad \text{for all } v^h \in V_0^h, \tag{4.3}$$

Let us note first that this method is conformal, which is due to the choice we made for the space V^h, for the trial function u^h, and the space V_0^h, for the test functions v^h. It is not strictly speaking consistent, because the boundary condition involving g^h is not necessarily verified for u. In fact

$$\Upsilon u = g^h + (g - g^h),$$

and the consistency error term, $(g - g^h)$, comes from Lagrange interpolation. Nevertheless, if we choose $v^h \in V_0^h \subset V_0$ instead of v in Problem (2.13), and subtract the equation with (4.3), we obtain a Galerkin orthogonality:

$$a(u - u^h, v^h) = 0, \quad \forall v^h \in V_0^h. \tag{4.4}$$

As $a(\cdot, \cdot)$ defines an inner product on V_0 (see Chap. 2), and since $V_0^h \subset V^h$, Galerkin orthogonality means that u^h is the orthogonal projection of u onto V_0^h. As a result, it is also the closest function to u in V_0^h in the sense of the norm induced by $a(\cdot, \cdot)$. The a priori error estimate will allow us to quantify this more precisely (and Galerkin orthogonality plays a fundamental role in this). Let us see now the mathematical analysis and the implementation aspects of this method.

4.2.1   Well-Posedness

The first thing to assess is that Problem (4.3) can be solved, in other terms that it admits one unique solution u^h, which corresponds (almost) to the solution a computer will deliver for a practical problem. This is the object of the following proposition:

Proposition 4.1 *The discrete problem* (4.3) *admits one unique solution u^h in V^h.*

Proof Let v_g be the lifting associated to the boundary data g. Since we supposed v_g continuous, we can take its Lagrange interpolant $\mathscr{I}^h(v_g)$. As in the proof of Theorem 2.7, for the continuous problem, we introduce the auxilliary problem:

Find $u_0^h \in V_0^h$ such that
$$a(u_0^h, v^h) = L(v^h) - a(\mathscr{I}^h(v_g), v^h) \quad \text{for all } v^h \in V_0^h. \tag{4.5}$$

Since $V_0^h \subset V_0$, $a(\cdot, \cdot)$ is still coercive on V_0^h, owing to Theorem 2.6. It results that there exists one unique solution u_0^h to Problem (4.5). Now thanks to the property (3.9), there holds:

$$\Upsilon(u_0^h + \mathscr{I}^h(v_g)) = \underbrace{\Upsilon u_0^h}_{=0} + \Upsilon(\mathscr{I}^h(v_g)) = \mathscr{I}_\Gamma^h(\Upsilon v_g) = \mathscr{I}_\Gamma^h(g) = g^h.$$

Furthermore, for any $v^h \in V_0^h$, Problem (4.5) can be reformulated as

$$a(u_0^h + \mathscr{I}^h(v_g), v^h) = a(u_0^h, v^h) + a(\mathscr{I}^h(v_g), v^h) = L(v^h),$$

so $u_0^h + \mathscr{I}^h(v_g)(= u^h) \in V^h$ is a solution to Problem (4.3). We proceed as in Theorem 2.7 to assess that this solution is unique.   □

As for the continuous case, Problem (4.3) can be reformulated as a minimization problem. We state this result below, the proof of which is straightforward.

Proposition 4.2 *Let us define the quadratic functional*

$$\mathscr{J}^h : V^h \ni v^h \mapsto \frac{1}{2}a(v^h, v^h) - L(v^h) \in \mathbb{R}. \tag{4.6}$$

Then this functional is Fréchet-differentiable up to any order. It is also strongly convex in V_0^h. As a result, it has a unique minimizer in the affine space

$$V_g^h := \{v^h \in V^h \mid \Upsilon v^h = g^h\},$$

which is the unique solution $u^h \in V^h$ to Problem (4.3).

4.2.2 A Priori Error Estimate

Let us now establish that the above solution u^h converges towards the exact solution u when the mesh is uniformly refined, in other terms, when h tends to 0. The easiest error estimates to obtain are in the H^1-norm, since they derive directly from the variational structure of our boundary value problem. Let us first derive an abstract error estimate, in the sense that it does not take into account the precise definition of V^h, that can be any conforming subspace of V. It is a Strang-type Lemma (Cea's Lemma is of no use here, since the method is not fully consistent).

Theorem 4.1 *Let $u \in V$ be the solution to Problem (2.13). The solution $u^h \in V^h$ to Problem (4.3) verifies the* a priori *error bound below:*

$$\|u - u^h\|_{1,\Omega} \le C \inf_{\substack{v^h \in V^h \\ \Upsilon v^h = g^h}} \|u - v^h\|_{1,\Omega}. \tag{4.7}$$

The constant $C > 0$ is independent of h and u. It is also independent of the physical parameter μ.

Proof We consider u, solution to Problem (2.13), and u^h, solution to Problem (4.3). We choose any $v^h \in V^h$ provided its trace Υv^h be equal to g^h. We apply first a triangular inequality, and then we use the coercivity of $a(\cdot, \cdot)$ on V_0 followed by the Galerkin orthogonality (4.4), and get:

$$
\begin{aligned}
\|u - u^h\|_{1,\Omega} &\leq \|u - v^h\|_{1,\Omega} + \|v^h - u^h\|_{1,\Omega} \\
&\leq \|u - v^h\|_{1,\Omega} + \frac{1}{\alpha\mu} \frac{a(v^h - u^h, v^h - u^h)}{\|v^h - u^h\|_{1,\Omega}} \\
&\leq \|u - v^h\|_{1,\Omega} + \frac{1}{\alpha\mu} \frac{a(v^h - u^h, v^h - u)}{\|v^h - u^h\|_{1,\Omega}},
\end{aligned}
$$

with $\alpha > 0$ the coercivity constant of $a(\cdot, \cdot)$ on V_0. We note that the first line allows to separate the interpolation error (caused by the approximation of V by V^h) from the discrete error (caused by the approximation of the continuous problem (2.13) by the discrete problem (4.3)). The assumption $\Upsilon v^h = g^h$ is important to ensure that

$$
v^h - u^h \in V_0^h,
$$

which allows the bound in the second line. Then the continuity of $a(\cdot, \cdot)$ on V yields

$$
\frac{a(v^h - u^h, v^h - u)}{\|v^h - u^h\|_{1,\Omega}} \leq \mu\|v^h - u\|_{1,\Omega}.
$$

We combine this inequality with the previous one to obtain

$$
\|u - u^h\|_{1,\Omega} \leq \left(1 + \frac{1}{\alpha}\right)\|u - v^h\|_{1,\Omega}.
$$

To obtain the estimation (4.7), it suffices to take the infimum over all the possible v^h.   □

Now for Lagrange finite elements, we can provide an *a priori* error estimate in the H^1-norm with an explicit convergence rate:

Corollary 4.1 *We suppose that $u \in H^2(\Omega)$. The solution u^h to Problem (4.3) satisfies*

$$
\|u - u^h\|_{1,\Omega} \leq Ch\|u\|_{2,\Omega}, \tag{4.8}
$$

where $C > 0$ is a constant independent of h and of u. The constant $C > 0$ is also independent of the physical parameter $\mu > 0$.

Proof The assumption $u \in H^2(\Omega)$ allows to ensure that u is continuous in $\overline{\Omega}$, even for $d = 3$, using Sobolev embeddings. As a result, we can define $\mathscr{I}^h u$, its Lagrange interpolant. Furthermore, thanks to property (3.9), there holds

$$\Upsilon(\mathscr{I}^h u) = \mathscr{I}_\Gamma^h(\Upsilon u) = \mathscr{I}_\Gamma^h(g) = g^h,$$

so we can take $v^h = \mathscr{I}^h u$ in the abstract error estimate (4.7). Furthermore, the local Lagrange interpolation estimate of Theorem 3.1, after summation on all the simplices $T \in \mathcal{T}^h$, yields:

$$\|u - \mathscr{I}^h u\|_{1,\Omega} \leq Ch\|u\|_{2,\Omega}.$$

We conclude using (4.7) in combination with the above estimation.   □

We obtain an optimal error bound in the above result, in the sense that the convergence rates are the same as if we would be capable of interpolating directly u using its value for the Lagrange nodes associated to V^h.

Finally, we can say something about the error on the non-homogeneous Dirichlet boundary condition. Let us suppose that $g \in H^1(\Gamma) \cap \mathscr{C}^0(\Gamma)$. Using the condition $\Upsilon u^h = g^h$, we can ensure that

$$\|\Upsilon u - \Upsilon u^h\|_{0,\Gamma} = \|g - g^h\|_{0,\Gamma} = \|g - \mathscr{I}_\Gamma^h(g)\|_{0,\Gamma} \leq Ch\|g\|_{1,\Gamma},$$

where we used Theorem 3.1. This is expected, since the boundary data is interpolated: we recover simply the interpolation error of g.

4.3   Implementation Aspects

This section details how the discrete problem (4.3) can be recasted as a linear system. At this stage, a computer can be used to solve it.

4.3.1   The Matrix System

Let us detail now how the finite element discretization (4.3) can be implemented for practical problems. We use the notations presented in the previous chapter (Chap. 3). We define $u_i := u^h(a_i)$, the value of the solution u^h at node a_i, $i = 1, \ldots, N$, and we collect all these degrees of freedom in a column vector U:

$$
U := \begin{bmatrix} u_1 \\ \vdots \\ u_{N_{\text{int}}} \\ \hline u_{N_{\text{int}}+1} \\ \vdots \\ u_N \end{bmatrix} =: \begin{bmatrix} U_{\text{I}} \\ \hline U_{\text{B}} \end{bmatrix},
$$

where the column vectors U_{I} and U_{B} are associated with the interior degrees of freedom and the boundary degrees of freedom, respectively.

Let us deal first with the boundary condition $\Upsilon u^h = g^h$ in (4.3). This implies that for every node a_i on the boundary, we obtain the corresponding degree of freedom u_i as follows:

$$
u_i = u^h(a_i) = g^h(a_i) = \mathscr{I}_\Gamma^h g(a_i) = g(a_i).
$$

Conversely, if we ensure $u_i = g(a_i)$ for every node i within the boundary set B, we get $\Upsilon u^h = g^h$, since the Lagrange interpolant is uniquely determined by the nodal values. In other terms, all the degrees of freedom on the Dirichlet boundary are obtained directly through evaluation of the boundary datum g at nodes a_i, for $i \in$ B.

Then the second equation of (4.3) can be written, if we choose, for any index $i \in$ I, the basis function φ_i as a test function ($v^h = \varphi_i \in V_0^h$):

$$
\sum_{j=1,\dots N} K_{ij} u_j = f_i, \quad \forall i \in \text{I},
$$

using the usual conventions

$$
K_{ij} := \mu \int_\Omega \nabla \varphi_i \cdot \nabla \varphi_j, \quad 1 \le i, j \le N.
$$

We split the sum into two parts; to differentiate interior nodes from boundary nodes, use the relationship $u_j = g(a_j)$ for the boundary nodes and get:

$$
\sum_{j \in \text{I}} K_{ij} u_j = f_i - \sum_{j \in \text{B}} K_{ij} u_j = f_i - \sum_{j \in \text{B}} K_{ij} g(a_j), \quad \forall i \in \text{I}.
$$

This can be expressed as below, in matrix form:

$$K_{\mathrm{I}}U_{\mathrm{I}} = F - K_{\mathrm{BI}}G,$$

where K_{I} is the stiffness submatrix associated solely to internal nodes:

$$K_{\mathrm{I},ij} = \int_{\Omega} \nabla\varphi_i \cdot \nabla\varphi_j, \qquad i, j \in \mathrm{I},$$

and K_{BI} is the submatrix that couples boundary and interior nodes:

$$K_{\mathrm{BI},ij} = \int_{\Omega} \nabla\varphi_i \cdot \nabla\varphi_j, \qquad i \in \mathrm{I},\ j \in \mathrm{B}.$$

The notations F et G are column vectors that contain, respectively, the components f_i from the source term f, for $i \in \mathrm{I}$, and the components $g(a_j)$ for the boundary data g, for $j \in \mathrm{B}$. To summarize, Problem (4.3) in matricial form can be written as follows:

$$K_{\mathrm{I}}U_{\mathrm{I}} = F - K_{\mathrm{BI}}G, \qquad U_{\mathrm{B}} = G.$$

In the engineering litterature, this method, in matricial form, is sometimes called a partitioning procedure. Note at this stage that, because of the symmetry and the coercivity of $a(\cdot,\cdot)$ on the discrete space V_0^h, the matrix K_{I} is symmetric positive definite. Also, because the finite element basis functions (φ_i) have their support localized in patches of simplices, the matrix K_{I} has also a sparse structure. As a result, the solution U can be obtained using an iterative solver such as a conjugate gradient method, though, for problems of small size, direct solvers such as Cholesky are employed [10, 125].

An equivalent global matrix formulation is depicted below:

$$\left[\begin{array}{c|c} K_{\mathrm{I}} & K_{\mathrm{BI}} \\ \hline 0 & I \end{array}\right] \left[\begin{array}{c} U_{\mathrm{I}} \\ \hline U_{\mathrm{B}} \end{array}\right] = \left[\begin{array}{c} F \\ \hline G \end{array}\right].$$

Remark that this formulation can be obtained directly from the global stiffness matrix K and right-hand side F, after an assembly procedure that does not distinguish between interior nodes and boundary nodes (it is sometimes simpler to proceed like this): it suffices to apply a postprocessing that replaces the corresponding lines of K and F, by the lines of the above system for the boundary nodes. This procedure is sometimes reefered to as an elimination procedure.

Remark 4.1 Let v_g^h be a simple discrete lifting of g^h, built as follows:

$$v_g^h(a_i) = g^h(a_i), i \in \mathrm{B}, \quad v_g^h(a_i) = 0, i \in \mathrm{I}.$$

Therefore, it is the unique function in V^h equal to g^h on the boundary Γ and equal to 0 for all the internal nodes (it is not equal to 0 only within the first layer of elements near the boundary). If we use v_g^h to build an equivalent homogeneous Dirichlet problem (4.5), we obtain the same matrix form as above [125, §3.2.2. pp.124–126, Remark 8.17 p.378].

4.3.2   Theory vs. Practice: The Numerical Errors

After implementation and solution with a computer, the vector $\tilde{U}$ of nodal unknowns delivered by the computer may differ slightly from the (theoretical, exact) vector U mentioned above and that serves to reconstruct u^h. The difference between $\tilde{U}$ and U is called the numerical error.

In this simple situation, it comes mostly from roundoff errors, and, possibly, from the numerical algorithm to solve the linear system. Particularly, iterative Krylov solvers like conjugate gradient are stopped after a few iterations [10]. Then, they generally deliver an accurate solution, but that may differ from the solution obtained at convergence.

More generally, another important source of numerical errors may come from iterative solvers to solve non-linear systems, if the boundary value problem involves non-linear operators (this is not the case here, but it will be in Chap. 6), and from numerical integration to compute the coefficients of K and F in the linear system. For simple low-order Lagrange finite elements on simplices, numerical integration (quadrature) errors can be avoided easily [8], and this is a benefit of the method, but this may not be the case for more general finite elements [87, 125, 249].

4.4   Pure Neumann Boundary Conditions

We sketch in this section how the pure Neumann boundary value problem (2.22) presented in Chap. 2 can be approximated and solved. We follow mostly P. Bochev and R.B. Lehoucq [51].

4.4.1   Discrete Neumann Problem

A discrete counterpart of Problem (2.22) reads

Find $u^h \in V^h$ that satisfies

$$a(u^h, v^h) = L_n(v^h) \quad \text{for all } v^h \in V^h, \tag{4.9}$$

If the compatibility condition is not satisfied, this problem has no solution, and if the compatibility condition holds, it has a set of solutions $\{u_m^h + C1_\Omega \mid C \in \mathbb{R}\}$, with u_m^h, the solution with average equal to zero. Moreover, the following error bound can be established, for u, any solution to (2.22) and u^h, any solution to (4.9):

$$\|P(u - u^h)\|_{1,\Omega} \leq Ch\|u\|_{2,\Omega}, \tag{4.10}$$

where $C > 0$ is a constant independent of h and of u, and P is the projection operator introduced in Chap. 2.

4.4.2   Implementation Aspects

We still note $u_i := u^h(a_i)$, the value of the solution u^h at node a_i, $i = 1, \ldots, N$, and we collect all these degrees of freedom in a column vector U. Still we keep the conventions

$$K_{ij} := \mu \int_\Omega \nabla\varphi_i \cdot \nabla\varphi_j, \qquad f_i := \int_\Omega f\varphi_i, \qquad 1 \leq i, j \leq N.$$

Problem (4.9) in matrix form reads

$$KU = F.$$

Since the matrix K is symmetric and positive semi-definite, the conjugate gradient method still works and delivers one solution among those admissible; see [10, 51]. However, roundoff errors can interfere with the compatibility condition at the discrete level [51]. A popular technique to get one unique solution is to impose a value for u^h at one node a_i of the mesh. See [51] for various variationally consistent techniques that are thoroughly studied.

4.5   A Numerical Illustration

The numerical test is done with scikit-fem [159]. The domain Ω is the unit square: $\Omega : (0, 1)^2$. We consider a very simple test case with the closed-form solution

$$u(x, y) = \sin(\pi x) \sin(\pi y)$$

on Ω. This corresponds to a boundary term

$$g = 0$$

on Γ and to a source term

$$f(x, y) = -\Delta u(x, y) = 2\pi^2 \sin(\pi x) \sin(\pi y). \tag{4.11}$$

Fig. 4.2 Magnitude of the finite element solution u^h to Poisson's problem with Dirichlet boundary conditions using Lagrange finite elements of order 1, a structured mesh and scikit-fem

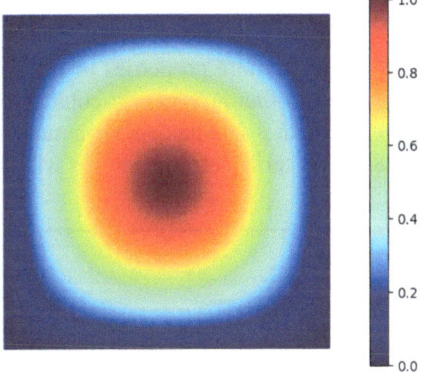

The solution to Problem 4.3 using a structured mesh is depicted in Fig. 4.2. The corresponding errors in L^2-norm and H^1-norm are of 3.10^{-4} and 5.10^{-3}, respectively. The mesh is relatively coarse, and this is to illustrate that the discrete error is not negligible, even in this very simple situation. Remark that for this test case, it is difficult to see the effect of numerical errors (for a deeper insight into this topic, see [215] and the numerical studies it presents).

4.6 Further Comments

This section presents some topics that have not been treated and related to the discretization of Dirichlet or Neumann boundary conditions.

4.6.1 Other Elliptic Operators and Boundary Conditions

A systematic study of the approximation of a scalar elliptic partial differential equation with different boundary conditions (Dirichlet, Neumann, mixed, and Fourier-Robin) is carried out in e.g. [125, 128].

4.6.2 Alternative Discretization Methods for Essential Boundary Conditions

Other common possibilities to approximate an essential boundary conditions are mixed methods and penalty methods and have been studied in two seminal papers of I. Babuska [25, 26]. Nitsche method [209], which is a special kind of mixed method, is detailed in the next chapter. For some surveys on the topic, see [78, 237].

4.6.3 Further Insight into the Mathematical Analysis

In fact, much more properties of the numerical methods can be established: error estimates in other norms (particularly the L^2-norm, with the Aubin-Nitsche duality trick), estimates for the condition number [125], and a discrete maximum principle [128]. A posteriori error estimates will be presented in the last chapter of this book.

Problems

Here we study the numerical approximation of an inclusion. This series of exercices is inspired from [9]. Let Ω be an open bounded set of $\mathbb{R}^2$ of boundary $\Gamma := \partial\Omega$, and that contains an inclusion K as a subdomain of Ω; see Fig. 4.3. We assume that K is a connected compact subset of $\mathbb{R}^2$. Also, we suppose that the boundaries Γ and ∂K are polygonal. Let n be the outward unit normal to the set $\Omega \setminus K$, for the external boundary Γ and on the boundary ∂K of the inclusion. Let $f : \Omega \setminus K \to \mathbb{R}$ be a source term (with $f \in L^2(\Omega \setminus K)$). We want to solve the following elliptic boundary value problem:

$$\begin{cases} -\Delta u = f \text{ in } \Omega \setminus K, \\ u = 0 \ \text{ on } \Gamma, \\ u = C \text{ on } \partial K, \\ \displaystyle\int_{\partial K} \frac{\partial u}{\partial n} = 0, \end{cases} \qquad (4.12)$$

Let

$$L^1(\partial K) = \left\{ v : \partial K \to \mathbb{R} \ \middle|\ \int_{\partial K} |v| < +\infty \right\}.$$

Fig. 4.3 A domain Ω for the diffusion equation with an inclusion K

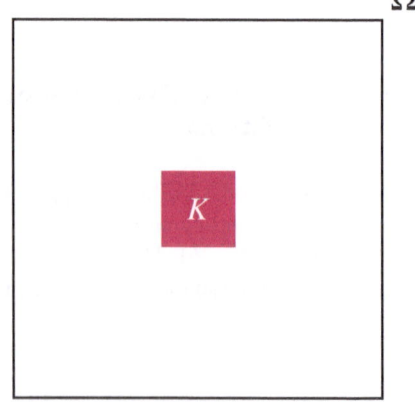

be the set of integrable functions on ∂K. Let $m : L^1(\partial K) \to \mathbb{R}$ that associates to each function $v \in L^1(\partial K)$, its mean value over ∂K:

$$m(v) = \frac{1}{|\partial K|} \int_{\partial K} v,$$

where $|\partial K|$ is the measure of the boundary ∂K. Let us define also

$$V := \{v \in H^1(\Omega \setminus K) \mid v = 0 \text{ on } \Gamma, \ v = m(v) \text{ on } \partial K\}.$$

4.1 Show that the condition $\exists C \in \mathbb{R}, \ u = C$ on ∂K is equivalent to $u = m(u)$ on ∂K.

4.2 Prove that V is a Hilbert space.

4.3 Prove that if u solves Problem (4.12), it also solves the weak form

$$a(u, v) = L(v), \quad \forall v \in V, \tag{4.13}$$

with

$$a(u, v) = \int_{\Omega \setminus K} \nabla u \cdot \nabla v, \qquad L(v) = \int_{\Omega \setminus K} f v.$$

4.4 Prove that if u is a solution to the weak form (4.13), it also satisfies all the equations of the strong Problem (4.12).

4.5 Prove that Problem (4.13) is well-posed.

4.6 Prove that the unique solution u to (4.13) is also the unique minimizer on V of a functional $J : V \to \mathbb{R}$ and provide the expression of this functional.

The formulation (4.13) is of interest for the theoretical study of the inclusion problem. However, it may be cumbersome for finite element approximation, notably because of the definition of the vector space V. So we introduce a penalty formulation, which consists in finding, over $H^1(\Omega \setminus K)$ (unconstrained vector space), the minimum of the functional $J_\varepsilon : H^1(\Omega \setminus K) \to \mathbb{R}$ whose expression is given below:

$$J_\varepsilon(v) := J(v) + \frac{1}{2\varepsilon} \int_\Gamma v^2 + \frac{1}{2\varepsilon} \int_{\partial K} (v - m(v))^2.$$

The real value $\varepsilon > 0$ is a penalty parameter.

4.7 Provide the expression of the weak formulation associated with the minimization of the penalized functional J_ε.

4.8 Prove that the weak formulation obtained at Problem 4.7 is well-posed, and that the solution is also the unique minimizer of J_ε on $H^1(\Omega \setminus K)$.

Let $\mathcal{T}_h$ be a simplicial mesh of the domain $\Omega \setminus K$, where h stands for the mesh size. We consider piecewise linear continuous Lagrange finite elements:

$$V_h = \{v_h \in \mathscr{C}^0(\overline{\Omega \setminus K}) \,|\, v_h|_T \in \mathbb{P}_1(T), \ \forall T \in \mathcal{T}_h\}.$$

4.9 Write the finite element formulation that is counterpart of the weak formulation of Problem 4.7. Prove that this finite element formulation is well-posed.

4.10 Detail the expression of the linear system associated with the finite element formulation of Problem 4.9. Prove that it is invertible.

Nitsche Finite Element Method

5

This chapter presents Nitsche method as a prototype of an alternative method to handle essential boundary conditions, and, moreover, as a prototype of non-standard finite element method, for which numerical analysis is helpful to fix some issues.

5.1 Outline

We are, as in the previous chapter, in between the second step of our pipeline and the third step; see Fig. 5.1.

5.1.1 An Alternative to the Standard Method

The standard method detailed in the previous chapter to handle essential boundary condition, though simple, has many limitations. In fact, it is fine when a nodal finite element method is used and for simple essential boundary conditions such as a Dirichlet boundary condition.

The first motivation to approximate differently essential boundary conditions concerns discrete variational techniques for which the degrees of freedom on the boundary are not nodal values, and in which it is not direct to approximate a Dirichlet condition. This is notably the case for unfitted finite elements or geometrically non-conforming finite elements, where the mesh boundary and the domain boundary do not match, as it occurs in fictitious domain methods, the eXtended finite element method (XFEM) or the cut finite element method; see e.g. [66, 111, 114, 153, 165, 199, 217]. It is also the case for isogeometric analysis (IGA) [95] and spline-based finite elements [123] where degrees of freedom are control points of B-spline functions and are not nodes on the boundary [33]. A similar situation occurs in domain decomposition techniques, in practical situations where it is easier to mesh independently subdomains. At the interfaces

© The Author(s), under exclusive license to Springer Nature Switzerland AG 2025
F. Chouly, *Finite Element Approximation of Boundary Value Problems*,
Compact Textbooks in Mathematics, https://doi.org/10.1007/978-3-031-72530-2_5

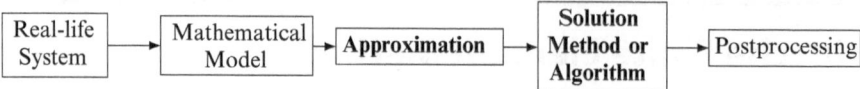

Fig. 5.1 The above pipeline depicts the global process behind a numerical simulation. We are ending with the second step of approximation and going to the third step

between subdomains, there are non-matching meshes [132, 221]. These cases can be treated with the mortar method [41,45,46,247] or Nitsche method [38,142,177]. Another motivation has been the design of more and more sophisticated numerical methods to solve efficiently problems involving more complex essential boundary or interface conditions. Just to mention a few ones with corresponding recent works, we can quote fluid-structure interaction [23, 134, 149], contact and friction [11,81,86,164,181,248], or even now fluid-structure contact where the contact conditions between elastic solids need to coexist with a viscous fluid between them [21,67,137,196,230].

5.1.2   The Original Nitsche Method

In its seminal paper, J.A. Nitsche introduced the following functional to approximate Poisson's problem with Dirichlet boundary conditions:

$$\mathcal{J}_N(v^h) = \mathcal{J}(v^h) - \int_\Gamma (v^h - g)(\mu \partial_n v^h) + \frac{1}{2}\int_\Gamma (v^h - g)^2$$

where $\mathcal{J}(v^h)$ is the quadratic functional associated with the continuous weak formulation (see Chap. 2), and $\gamma > 0$ is a positive function on the boundary. The corresponding weak form is obtained as the first-order optimality condition of the above functional and reads: find $u^h \in V^h$ solution to

$$a(u^h, v^h) - \int_\Gamma (\mu \partial_n u^h) v^h - \int_\Gamma (u^h - g)(\mu \partial_n v^h) + \int_\Gamma \gamma (u^h - g) v^h - L(v^h) = 0.$$

Remark the special term $-\int_\Gamma (v^h - g)(\mu \partial_n v^h)$ in the Nitsche functional that allows to preserve consistency and makes the method differ from standard penalization or regularization techniques.

5.1.3   The Role of the Stability and Convergence Analysis

The delicate issue in Nitsche method, as in stabilized mixed methods, is to choose in the right way the positive function γ on the boundary. First, why is it positive? Should it not be negative? Should it be chosen close to zero or is it better to set large values? Though ad hoc arguments can be used to fix this issue, here the mathematical analysis is particularly helpful to answer these questions. Let us introduce the following notation for the symmetric Nitsche bilinear form:

$$A_{N,1}(u^h, v^h) := a(u^h, v^h) - \int_\Gamma (\mu \partial_n u^h) v^h - \int_\Gamma u^h (\mu \partial_n v^h) + \int_\Gamma \gamma u^h v^h.$$

To simplify, let us consider a uniform or quasi-uniform mesh, and in this case the function γ can be chosen as a constant on the whole boundary:

$$\gamma = \frac{\gamma_0}{h},$$

where γ_0, the Nitsche parameter, needs to be fixed accordingly.

First, let us have a look at the stability of the method. Using simple and usual arguments such as Cauchy-Schwarz inequality, Young inequality and a discrete trace inequality, and assuming the Nitsche paramater γ_0 is positive and large enough, we can show that

$$A_{N,1}(v^h, v^h) \geq C \left(a(v^h, v^h) + \int_\Gamma |v^h|^2 \right)$$

for all $v^h \in V^h$ and $C > 0$, a given constant independent of h. This discrete ellipticity property ensures that the Nitsche formulation is well-posed and admits one unique solution u^h. Also, we can derive an a priori error estimate that implies that, for piecewise linear Lagrange finite elements, and a regular enough solution, there holds

$$\|u - u^h\|_{1,\Omega} \leq Ch, \tag{5.1}$$

where $C > 0$ that does not depend on h. In the proof, the appropriate scaling in $O(1/h)$ of the Nitsche term plays a fundamental role.

As a result, not only the stability and convergence analysis allows to ensure here that Nitsche method has no major pathology, but also it provides useful information to the user on the appropriate value of the Nitsche function γ: it needs to be positive, to scale as the inverse of the mesh size, and to be large enough.

5.1.4 Implementation

For the standard symmetric Nitsche method applied to Dirichlet boundary condition, there is no special issue relative to its implementation. Conversely to the standard method presented in Chap. 4, there is no special treatment in the assembly of the matrix related to the bulk bilinear form, with the separation of boundary nodes and interior nodes. Special data structures may be useful for the assembly of Nitsche terms, but this can be managed without much pain in most of the finite element codes. For the symmetric Nitsche method, a symmetric definite positive global matrix is obtained that can be solved with any appropriate direct or iterative standard method.

The method is implemented, for instance, in GetFEM [224], in FEniCS [16], or in scikit-fem [79], just to mention a few usual open-source finite element libraries.

5.2   A Simple Derivation of Nitsche Method

First, we describe heuristics to obtain the simplest version of Nitsche method that we will call incomplete Nitsche. Since this version lacks symmetry, we will see another way to recover symmetry using a modified energy functional. Then we will be able to derive a whole family of methods that depend on a real parameter θ that allows to switch between them. Last but not least, we present a penalty-free Nitsche method as recently revisited by E. Burman.

5.2.1   The Simplest Nitsche Method

Take $\gamma > 0$ a positive function on the boundary Γ. Let u be the solution to (2.8), and let v be a test function. Suppose that both functions are regular enough so that the following calculations make sense. From (2.8)–(i) and using Green formula (2.6) we get first

$$a(u, v) - \int_\Gamma \mu(\partial_n u)v = L(v).$$

Then we use the Dirichlet condition (2.8)–(ii) to write

$$\int_\Gamma \gamma(u - g)v = 0.$$

Now we add the two above equations:

$$a(u, v) - \int_\Gamma \mu(\partial_n u)v + \int_\Gamma \gamma uv = L(v) + \int_\Gamma \gamma gv. \qquad (5.2)$$

The boundary term $-\int_\Gamma \mu(\partial_n u)v$ above allows to recover consistency, as we will prove later on. The above weak form may have no meaning at the continuous level but has a well-defined discrete counterpart. For this purpose, let us take γ_N, a positive constant: the Nitsche parameter.

Let $\mathcal{T}^h$ be a given mesh of Ω, and we use the terminology *facet* to design either an edge in two dimesions or a face in three dimensions. We call a generic facet F. Then define locally in the interior of each boundary facet $F \subset \Gamma$

$$\gamma|_F = \mu\gamma_N h_F^{-1}, \qquad (5.3)$$

with $|_F$ that designs, in fact, the restriction to the interior of F. This is a slight abuse of notation and, in fact, the value of γ at the intersection between two facets is of no importance and can be set to 0, for instance. The notation h_F stands for the diameter of the facet F.

We use the Lagrange finite element space V^h of Chap. 3 (no condition on the boundary is directly imposed on the finite elements, conversely to the standard method that involved V_0^h). We take the trial function u^h and the test function v^h that belong to V^h. To lighten the writing of the weak form (5.2), we introduce the (non-symmetric) bilinear form

$$A_{N,0}(u^h, v^h) := a(u^h, v^h) - \int_\Gamma \mu(\partial_n u^h)v^h + \int_\Gamma \gamma u^h v^h$$

and the linear form

$$l_{N,0}(v^h) := L(v^h) + \int_\Gamma \gamma g v^h.$$

The incomplete Nitsche method for Poisson's problem (2.8) reads

$$\begin{cases} \text{Find } u^h \in V^h \text{ solution to} \\ A_{N,0}(u^h, v^h) = l_{N,0}(v^h), \quad \forall v^h \in V^h. \end{cases} \tag{5.4}$$

5.2.2 Recovering Symmetry

To recover symmetry, we can follow the path of J.A. Nitsche original formulation [209] and obtain a symmetric Nitsche method as the first-order optimality condition associated to the modified energy functional

$$\mathcal{J}_N : V^h \ni v^h \mapsto \mathcal{J}(v^h) - \int_\Gamma (v^h - g)(\mu \partial_n v^h) + \frac{1}{2}\int_\Gamma (v^h - g)^2 \in \mathbb{R}.$$

We compute $\mathcal{J}'_N$:

$$\mathcal{J}'_N(u^h; v^h)$$

$$= a(u^h, v^h) - L(v^h) - \int_\Gamma (\mu \partial_n u^h)v^h - \int_\Gamma (u^h - g)(\mu \partial_n v^h) + \int_\Gamma \gamma(u^h - g)v^h.$$

The first-order optimality condition $\mathcal{J}'_N(u^h; \cdot) = 0$ yields the symmetric Nitsche method:

$$\begin{cases} \text{Find } u^h \in V^h \text{ solution to} \\ A_{N,1}(u^h, v^h) = l_{N,1}(v^h), \quad \forall v^h \in V^h. \end{cases} \tag{5.5}$$

In the above formulation, the Nitsche bilinear form is

$$A_{N,1}(u^h, v^h) := a(u^h, v^h) - \int_\Gamma (\mu \partial_n u^h) v^h - \int_\Gamma u^h (\mu \partial_n v^h) + \int_\Gamma \gamma u^h v^h$$

and the corresponding linear form is

$$l_{N,1}(v^h) := L(v^h) + \int_\Gamma g(\gamma v^h - \mu \partial_n v^h).$$

5.2.3   Many Variants

Now, let us do something a bit more general, and we will see if it allows to recover in a unified form the two previous Nitsche methods. As a by-product, we will get a skewsymmetric variant. We start with $\theta \in \mathbb{R}$ a fixed parameter. Still, let u be the solution to (2.8), and v be a test function. Still, suppose that both functions are regular enough so that the following calculations make sense. From (2.8)–(i) and using Green formula (2.6), we still have

$$a(u, v) - \int_\Gamma (\mu \partial_n u) v = L(v).$$

Then we use the Dirichlet condition in a different form, to write

$$\int_\Gamma u(\gamma v - \theta \mu \partial_n v) = \int_\Gamma g(\gamma v - \theta \mu \partial_n v).$$

We sum the two above equations:

$$a(u, v) - \int_\Gamma (\mu \partial_n u) v - \theta \int_\Gamma u(\mu \partial_n v) + \int_\Gamma \gamma u v$$
$$= L(v) + \int_\Gamma g(\gamma v - \theta \mu \partial_n v). \tag{5.6}$$

As for the incomplete variant, a discrete formulation is obtained by setting γ as a piecewise constant function on the boundary Γ, using Eq. (5.3). We introduce the bilinear form

$$A_{N,\theta}(u^h, v^h) := a(u^h, v^h) - \int_\Gamma (\mu \partial_n u^h) v^h - \theta \int_\Gamma u^h (\mu \partial_n v^h) + \int_\Gamma \gamma u^h v^h$$

and the linear form

$$l_{N,\theta}(v^h) := L(v^h) + \int_\Gamma g(\gamma v^h - \theta \mu \partial_n v^h).$$

As a result, a family of Nitsche methods, indexed by θ, for Poisson's problem, with Dirichlet boundary conditions reads

$$\begin{cases} \text{Find } u^h \in V^h \text{ solution to} \\ A_{N,\theta}(u^h, v^h) = l_{N,\theta}(v^h), \quad \forall\, v^h \in V^h. \end{cases} \tag{5.7}$$

Three notable variants of the method for different values of the parameter θ can be obtained, as for the discontinuous Galerkin interior penalty (dGIP) method [109]:

1. For $\theta = 1$, we recover the symmetric variant just presented in the above Sect. 5.2.2.
2. For $\theta = 0$, we get the simplest, incomplete, formulation, presented at the beginning in 5.2.1. See also, for instance, [128, Section 37.1].
3. For $\theta = -1$, we recover the skew-symmetric formulation of J. Freund and R. Stenberg [140] (see also [65]).

5.2.4   Penalty-Free Nitsche

Let us now focus on the skewsymmetric variant $\theta = -1$. It reads

$$a(u^h, v^h) - \int_\Gamma (\mu \partial_n u^h) v^h + \int_\Gamma u^h (\mu \partial_n v^h) + \int_\Gamma \gamma u^h v^h$$
$$= \int_\Omega f v^h + \int_\Gamma g(\gamma v^h + \mu \partial_n v^h),$$

for $v^h \in V^h$ As noticed first by J. Freund and R. Stenberg [140], and then by E. Burman [65], it still performs well when the Nitsche parameter γ_N is set to zero and then reads

$$a(u^h, v^h) - \int_\Gamma (\mu \partial_n u^h) v^h + \int_\Gamma u^h (\mu \partial_n v^h) = \int_\Omega f v^h + \int_\Gamma g(\mu \partial_n v^h).$$

This means the Dirichlet boundary condition is imposed weakly thanks to

$$\int_\Gamma u^h (\mu \partial_n v^h) = \int_\Gamma g(\mu \partial_n v^h)$$

for $v^h \in V^h$. If one accepts to lose the symmetry, this method is very attractive in practice: 1) it remains consistent and implies no regularization; 2) it involves no extra unknown; and 3) it is parameter-free.

5.3   Well-Posedness and A Priori Error Bound

First, we provide a discrete trace inequality that plays a central role in the analysis. Then we establish a fundamental consistency result, which is at the origin of the performance of the method. Then we detail the well-posedness of Nitsche method and present an a priori error bound in the natural norm. We try to underline the condition this requires on the Nitsche parameter γ_N.

5.3.1   A Discrete Trace Inequality

For the analysis of Nitsche method, it will be convenient to make use of the discrete norms introduced below. For $v \in L^2(\Gamma)$, we define

$$\|v\|^2_{-1/2,h,\Gamma} := \sum_{F \subset \Gamma} h_F \|v\|^2_{0,F}, \qquad \|v\|^2_{1/2,h,\Gamma} := \sum_{F \subset \Gamma} h_F^{-1} \|v\|^2_{0,F}.$$

For $v \in H^1(\Omega)$, we set

$$\|v\|_h := \left(\|\nabla v\|^2_{0,\Omega} + \|v\|^2_{1/2,h,\Gamma} \right)^{\frac{1}{2}}$$

which is an equivalent norm of the $H^1(\Omega)$-norm. Particularly, we can show the following:

Proposition 5.1 *There exists a constant $c > 0$, independent from h, such that for an arbitrary $v \in V$, there holds*

$$c\|v\|_{1,\Omega} \leq \|v\|_h.$$

Proof This is a direct consequence of the Deny-Lions Theorem (Theorem 2.8).   □

The following discrete trace inequality (often called discrete trace inverse inequality) will be needed:

Lemma 5.1 *Let V^h be the piecewise linear Lagrange finite element space as introduced in Chap. 3, built upon a shape-regular mesh. Then there exists a constant $c_I > 0$ independent of the mesh size h such that for any $v^h \in V^h$:*

$$\left\| \partial_n v^h \right\|^2_{-1/2,h,\Gamma} \leq c_I \|\nabla v^h\|^2_{0,\Omega}. \tag{5.8}$$

The constant c_I only depends on the shape regularity of the mesh and of the dimension d.

Proof Let F be a boundary facet that belongs to Γ and let T be its corresponding simplex. Take v^h in V^h. Since the gradient of v^h has a constant value in T and the same constant value in F, we have

$$\|\partial_n v^h\|_{0,F} \leq \|\nabla v^h\|_{0,F}$$

$$= \frac{|F|^{\frac{1}{2}}}{|T|^{\frac{1}{2}}}\|\nabla v^h\|_{0,T}$$

We now use

$$|F| \leq h_F^{d-1}, \qquad |T| \geq c_\rho h_T^d, \qquad \frac{h_F}{h_T} \leq c_\rho$$

where $c_\rho > 0$ depends solely on the shape regularity of the mesh. Then we get

$$\|\partial_n v^h\|_{0,F} \leq c_I h_F^{\frac{d-1}{2}} h_T^{-\frac{d}{2}} \|\nabla v^h\|_{0,T}$$

$$\leq c_I h_F^{-\frac{1}{2}} \|\nabla v^h\|_{0,T}.$$

By summation on all the edges $F \subset \Gamma$, we get (5.8). $\square$

Remark 5.1 For a proof with Lagrange finite elements of arbitrary order k, see, for instance, [86, Lemma 4.1] or [127, Lemma 12.8] (or also [242, Lemma 2.1] when the mesh is quasi-uniform). An estimate of the constant c_I for arbitrary order k is established in [245] (see also [127, Lemma 12.2]).

5.3.2 Consistency

Let us show that the Nitsche method (5.7) is consistent:

Lemma 5.2 *Suppose that the solution u to (2.8) belongs to $H^2(\Omega)$, then u solves also:*

$$A_{N,\theta}(u, v^h) = L_{N,\theta}(v^h), \quad \forall\, v^h \in V^h.$$

Proof Take u as the solution to (2.8) and consider $v^h \in V^h$. Since $u \in H^2(\Omega)$, there holds $\partial_n u \in L^2(\Gamma)$ and $A_{N,\theta}(u, v^h)$ is meaningful. First, we use the Dirichlet boundary condition $u = g$ on Γ to write

$$A_{N,\theta}(u, v^h) = a(u, v^h) - \int_\Gamma \mu \partial_n u v^h - \theta \int_\Gamma u \mu \partial_n v^h + \int_\Gamma \gamma_N u v^h$$

$$= a(u, v^h) - \int_\Gamma \mu \partial_n u v^h - \theta \int_\Gamma g \mu \partial_n v^h + \int_\Gamma \gamma_N g v^h$$

$$= a(u, v^h) - \int_\Gamma \mu \partial_n u v^h + \int_\Gamma g(\gamma_N v^h - \theta \mu \partial_n v^h).$$

From (2.8) and the Green formula, we get

$$a(u, v^h) - \int_\Gamma \mu \partial_n u v^h = L(v^h).$$

We combine this equality with the previous one and the proof is finished.   □

5.3.3 Well-Posedness

Conversely to what happened until there, the bilinear form $A_{N,\theta}(\cdot, \cdot)$ in Nitsche formulation is non-symmetric when θ is different from 1. As a result, the Riesz-Fré Representation Theorem cannot be used here to ensure well-posedness at the discrete level for all the values of θ. We need a more powerful result, adapted to non-symmetric bilinear forms (for the proof, see, for instance, [60] or [9, 232]):

Theorem 5.1 (Lax-Milgram) *Let V be a Hilbert space and $a(\cdot, \cdot)$ a continuous bilinear form, coercive on V. Let $L(\cdot)$ be a linear form continuous on V. Then the problem find $u \in V$ such that*

$$a(u, v) = L(v), \quad \forall v \in V,$$

admits one unique solution $u \in V$.

We show now that the Nitsche formulation is well-posed provided the Nitsche parameter is taken large enough.

Theorem 5.2 *Let $0 < \alpha_N < 1$ be a given constant that can be chosen arbitrarily and independently of h, μ, θ, and γ_0. Suppose that*

$$\gamma_N \geq (1 + \theta)^2 \frac{c_I}{4(1 - \alpha_N)^2}, \tag{5.9}$$

where $c_I > 0$ is the trace-inverse constant in (5.8). Then, for $v^h \in V^h$, we have:

$$A_{N,\theta}(v^h, v^h) \geq \mu \alpha_N \min(1, \gamma_N) \|v^h\|_h^2. \tag{5.10}$$

Moreover, for any value of $\gamma_N > 0$ that satisfies also condition (5.9), the Nitsche formulation (5.7) admits one unique solution $u^h \in V^h$ that verifies the bound

$$\|u^h\|_h \leq \frac{1}{\alpha_N \min(1, \gamma_N)} \left(\frac{C}{\mu} \|f\|_{0,\Omega} + \left(\gamma_N + |\theta| c_I^{\frac{1}{2}} \right) \|g\|_{1/2,h,\Gamma} \right),$$

with $C > 0$ a constant that comes from Proposition 5.1.

Proof Let us take v^h in V^h. Using Cauchy-Schwarz and Young ($ab \leq \beta a^2/2 + b^2/(2\beta)$, with $\beta > 0$ arbitrary), we get

$$A_{N,\theta}(v^h, v^h)$$

$$= a(v^h, v^h) - (1+\theta) \int_\Gamma \mu \partial_n v^h v^h + \int_\Gamma \gamma v^h v^h$$

$$= \mu \|\nabla v^h\|_{0,\Omega}^2 - (1+\theta)\mu \int_\Gamma (\gamma^{-\frac{1}{2}} \partial_n v^h)(\gamma^{\frac{1}{2}} v^h) + \mu \gamma_N \left\| v^h \right\|_{1/2,h,\Gamma}^2$$

$$\geq \mu \|\nabla v^h\|_{0,\Omega}^2 - |1+\theta| \mu^{\frac{1}{2}} \|\gamma^{-\frac{1}{2}} \partial_n v^h\|_{0,\Gamma} \mu^{\frac{1}{2}} \|\gamma^{\frac{1}{2}} v^h\|_{0,\Gamma} + \mu \gamma_N \left\| v^h \right\|_{1/2,h,\Gamma}^2$$

$$\geq \mu \|\nabla v^h\|_{0,\Omega}^2 - (1+\theta)^2 \frac{\beta\mu}{2\gamma_N} \left\| \partial_n v^h \right\|_{-1/2,h,\Gamma}^2 + \mu \gamma_N \left(1 - \frac{1}{2\beta} \right) \left\| v^h \right\|_{1/2,h,\Gamma}^2.$$

We use now the discrete trace inequality (5.8) to bound the term involving the normal derivative:

$$A_{N,\theta}(v^h, v^h) \geq \mu \left(1 - (1+\theta)^2 \frac{\beta c_I}{2\gamma_N} \right) \|\nabla v^h\|_{0,\Omega}^2 + \mu \gamma_N \left(1 - \frac{1}{2\beta} \right) \left\| v^h \right\|_{1/2,h,\Gamma}^2.$$

Let us take $0 < \alpha_N < 1$ arbitrary and take now

$$\beta = \frac{1}{2(1-\alpha_N)}$$

such that

$$1 - \frac{1}{2\beta} = 1 - \frac{1}{2\frac{1}{2(1-\alpha_N)}} = \alpha_N.$$

As a result, there holds

$$A_{N,\theta}(v^h, v^h) \geq \mu \left(1 - (1+\theta)^2 \frac{c_I}{4(1-\alpha_N)\gamma_N} \right) \|\nabla v^h\|_{0,\Omega}^2 + \mu \gamma_N \alpha_N \left\| v^h \right\|_{1/2,h,\Gamma}^2.$$

Take now

$$\gamma_N \geq (1+\theta)^2 \frac{c_I}{4(1-\alpha_N)^2},$$

which is exactly condition (5.9). This implies

$$1 - (1+\theta)^2 \frac{c_I}{4(1-\alpha_N)\gamma_N} \geq \alpha_N.$$

Observe that this remains valid whatever the value of θ is (even for $\theta = -1$). We get finally

$$A_{N,\theta}(v^h, v^h) \geq \mu\alpha_N\|\nabla v^h\|_{0,\Omega}^2 + \mu\gamma_N\alpha_N \left\|v^h\right\|_{1/2,h,\Gamma}^2$$

$$\geq \mu\alpha_N \min(1, \gamma_N)\|v^h\|_h^2.$$

which is (5.10).

Suppose now that $\gamma_N > 0$ satisfies also the condition (5.9). Then the discrete coercivity condition (5.10) combined with the Lax-Milgram Theorem 5.1 ensures the well-posedness of the Nitsche formulation (5.7).

The a priori bound is obtained from (5.10) and (5.7) as follows: Let $u^h \in V^h$ be the solution to Problem (5.7). Then

$$\mu\alpha_N \min(1, \gamma_N)\|u^h\|_h^2 \leq A_{N,\theta}(u^h, u^h) = l_{N,\theta}(u^h),$$

and then we bound

$$|l_{N,\theta}(u^h)|$$

$$\leq \|f\|_{0,\Omega}\|u^h\|_{0,\Omega} + \mu\gamma_N\|g\|_{1/2,h,\Gamma}\left\|u^h\right\|_{1/2,h,\Gamma} + \mu|\theta|\|g\|_{1/2,h,\Gamma}\left\|\partial_n u^h\right\|_{-1/2,h,\Gamma}.$$

We use Proposition 5.1 to bound $\|f\|_{0,\Omega}$ and (5.8) to bound $\left\|\partial_n u^h\right\|_{-1/2,h,\Gamma}$, and get

$$|l_{N,\theta}(u^h)|$$

$$\leq C\|f\|_{0,\Omega}\|u^h\|_h + \mu\gamma_N\|g\|_{1/2,h,\Gamma}\left\|u^h\right\|_{1/2,h,\Gamma} + \mu|\theta|c_I^{\frac{1}{2}}\|g\|_{1/2,h,\Gamma}\|\nabla u\|_{0,\Omega},$$

with $C > 0$, a constant that comes from Proposition 5.1. We combine the previous estimates and get

$$\mu\alpha_N \min(1, \gamma_N)\|u^h\|_h \leq C\|f\|_{0,\Omega} + \mu\gamma_N\|g\|_{1/2,h,\Gamma} + \mu|\theta|c_I^{\frac{1}{2}}\|g\|_{1/2,h,\Gamma}.$$

We factorize the two last terms:

$$\mu\alpha_N \min(1, \gamma_N)\|u^h\|_h \leq C\|f\|_{0,\Omega} + \mu(\gamma_N + |\theta|c_I^{\frac{1}{2}})\|g\|_{1/2,h,\Gamma}.$$

We get finally

$$\|u^h\|_h \leq \frac{C}{\mu\alpha_N \min(1,\gamma_N)}\|f\|_{0,\Omega} + \frac{1}{\alpha_N \min(1,\gamma_N)}(\gamma_N + |\theta|c_I^{\frac{1}{2}})\|g\|_{1/2,h,\Gamma},$$

which yields

$$\|u^h\|_h \leq \frac{C}{\alpha_N \min(1,\gamma_N)}\left(\frac{1}{\mu}\|f\|_{0,\Omega} + (\gamma_N + |\theta|c_I^{\frac{1}{2}})\|g\|_{1/2,h,\Gamma}\right).$$

This ends the proof.

$\square$

Remark 5.2 There are some points in the above result that deserve some attention:

- It is not usual to have a discrete coercivity constant α_N that can be set arbirtrarily, but this allows to recover an almost optimal lower bound on the value of the Nitsche parameter γ_N (see below). In fact, there is an interplay between the value of the Nitsche parameter and discrete coercivity, in the sense that one can lower the value of Nitsche parameter if he accepts to weaken the coercivity condition.
- It is very easy from (5.9) and (5.10) to get a coercivity constant that does not depend at all on γ_N. For instance, when $\theta = 0$ we get

$$A_{N,0}(v^h, v^h) \geq \mu\alpha_N \min\left(1, \frac{c_I}{4(1-\alpha_N)^2}\right)\|v^h\|_h^2.$$

- In the case of the skew-symmetric version, we recover from (5.9) and (5.10) that well-posedness is ensured whatever the value of $\gamma_N > 0$ is, and a coercivity constant independent of γ_N can be recovered by simply supposing that $\gamma_N > \alpha_{-1}$, with $\alpha_{-1} > 0$, a small, arbitrary fixed, constant. For $\gamma_N = 0$ (penalty-free Nitsche), see the reference [65].
- In [128], a similar condition than (5.9) is recovered for the Nitsche parameter γ_N in the cases $\theta = 1$ and $\theta = 0$, and that correspond to the limit when α_N becomes very small. The authors proceed in a different way, using the inequality

$$x^2 - 2\beta xy + \gamma y^2 \geq \frac{\gamma - \beta^2}{1+\gamma}(x^2 + y^2),$$

that allows to recover a slightly better condition than (5.9) but at the price of a coercivity constant that depends of γ_N.
- For the symmetric version, the Riesz Representation Theorem would have been enough to ensure well-posedness, as in the continuous case. Nevertheless, in the general case, since the discretization with Nitsche breaks the symmetry, we need a stronger result of functional analysis, the Lax-Milgram lemma, to conclude

for well-posedness. Note that this is an interesting case of application of the Lax-Milgram lemma, since it is unexpected in some sense from the continuous problem.

5.3.4   Abstract A Priori Error Bound

Let us start with the following abstract error estimate:

Theorem 5.3 *Suppose that the Nitsche parameter $\gamma_N > 0$ satisfies (5.9). Let $u \in H^2(\Omega)$ be the solution to Problem (2.13) and u^h be the solution to Nitsche method (5.7). An a priori error estimate is given by*   .

$$\|u - u^h\|_h \tag{5.11}$$

$$\leq C \inf_{v^h \in V^h} \left(\|u - v^h\|_{1,\Omega} + \left\|u - v^h\right\|_{1/2,h,\Gamma} + \left\|\partial_n u - \partial_n v^h\right\|_{-1/2,h,\Gamma} \right).$$

The above constant $C > 0$ is independent of u and h.

Proof Suppose that $\gamma_N > 0$ verifies (5.9). Thus, previous Theorem 5.2 ensures the existence and uniqueness of a solution u^h. To simplify, we denote by

$$\alpha_A := \mu \alpha_N \min(1, \gamma_N)$$

the coercivity constant in Eq. (5.10). Take $u \in H^2(\Omega)$ the solution to Problem (2.13) and $u^h \in V^h$ the solution to Problem (5.7). Take an arbitrary discrete test function $v^h \in V^h$. First using a triangular inequality and then the V^h-ellipticity of $A_{N,\theta}(\cdot, \cdot)$, Eq. (5.10), we get

$$\|u - u^h\|_h \leq \|u - v^h\|_h + \|v^h - u^h\|_h$$

$$\leq \|u - v^h\|_h + \frac{1}{\alpha_A} \frac{A_{N,\theta}(v^h - u^h, v^h - u^h)}{\|v^h - u^h\|_h}.$$

Then, we use the consistency of the method, see Lemma 5.2, to write a Galerkin orthogonality:

$$A_{N,\theta}(u^h, v^h - u^h) = A_{N,\theta}(u, v^h - u^h).$$

This allows to get

$$\|u - u^h\|_h \leq \|u - v^h\|_h + \frac{1}{\alpha_A} \frac{A_{N,\theta}(v^h - u, v^h - u^h)}{\|v^h - u^h\|_h}.$$

Now using only Cauchy-Schwarz, we bound

$$A_{N,\theta}(v^h - u, v^h - u^h)$$

$$\leq \mu \|\nabla(v^h - u)\|_{0,\Omega} \|\nabla(v^h - u^h)\|_{0,\Omega} + \mu \left\| \partial_n v^h - \partial_n u \right\|_{-1/2,h,\Gamma} \left\| v^h - u^h \right\|_{1/2,h,\Gamma}$$

$$+ \mu |\theta| \left\| v^h - u \right\|_{1/2,h,\Gamma} \left\| \partial_n v^h - \partial_n u^h \right\|_{-1/2,h,\Gamma}$$

$$+ \mu \gamma_N \left\| v^h - u \right\|_{1/2,h,\Gamma} \left\| v^h - u^h \right\|_{1/2,h,\Gamma}.$$

We define

$$\mathcal{T}(v^h - u^h) := \|\nabla(v^h - u^h)\|_{0,\Omega} + \left\| \partial_n v^h - \partial_n u^h \right\|_{-1/2,h,\Gamma} + \left\| v^h - u^h \right\|_{1/2,h,\Gamma}$$

and simplify the expression of the bound $A_{N,\theta}(v^h - u, v^h - u^h)$ as

$$A_{N,\theta}(v^h - u, v^h - u^h)$$

$$\leq \mu \left(\|\nabla(v^h - u)\|_{0,\Omega} + \left\| \partial_n v^h - \partial_n u \right\|_{-1/2,h,\Gamma} + (|\theta| + \gamma_N) \left\| v^h - u \right\|_{1/2,h,\Gamma} \right)$$

$$\mathcal{T}(v^h - u^h).$$

Now the discrete trace inequality (5.8) allows to bound

$$\mathcal{T}(v^h - u^h) \leq C \|v^h - u^h\|_h$$

with C, a constant that depends only on c_I. As a result

$$\frac{A_{N,\theta}(v^h - u, v^h - u^h)}{\|v^h - u^h\|_h}$$

$$\leq C\mu \left(\|\nabla(v^h - u)\|_{0,\Omega} + \left\| \partial_n v^h - \partial_n u \right\|_{-1/2,h,\Gamma} + (|\theta| + \gamma_N) \left\| v^h - u \right\|_{1/2,h,\Gamma} \right).$$

There remains to notice that

$$\|u - v^h\|_h \leq C \left(\|\nabla(v^h - u)\|_{0,\Omega} + \left\| v^h - u \right\|_{1/2,h,\Gamma} \right),$$

in order to get

$$\|u - u^h\|_h \leq C \left(\|\nabla(v^h - u)\|_{0,\Omega} + \left\| \partial_n v^h - \partial_n u \right\|_{-1/2,h,\Gamma} + \left\| v^h - u \right\|_{1/2,h,\Gamma} \right),$$

where $C > 0$ does not depend on h nor on u. This gives (5.11).   □

Remark 5.3 The constant $C > 0$ in the above bound may depend on θ and γ_0. For a refined analysis, see, for instance, [78, 128].

5.3.5   Estimate for Low-Order Lagrange Elements

For low-order Lagrange finite elements on simplices, we can make explicit the convergence rate of the Nitsche finite element approximation as follows:

Corollary 5.1 *Suppose that the Nitsche parameter $\gamma_N > 0$ satisfies (5.9). Let $u \in H^2(\Omega)$ be the solution to Problem (2.13) and u^h be the solution to Nitsche method (5.7) discretized with piecewise linear Lagrange finite elements (V^h defined by (3.7)). The explicit* a priori *error estimate below holds:*

$$\|u - u^h\|_h \le Ch\|u\|_{2,\Omega}, \tag{5.12}$$

where the constant $C > 0$ is independent of u and h.

Proof Since $u \in H^2(\Omega)$, we can define $\mathscr{I}^h u$, its Lagrange interpolant. We choose $v^h = \mathscr{I}^h u \in V^h$ in the abstract estimate (5.11). The first two terms are estimated directly using Theorem 3.1. For the third term $\left\| \partial_n u - \partial_\mathbf{n}(\mathscr{I}^h u) \right\|_{-1/2,h,\Gamma}$, we proceed following [142]. We need the Scott-Zhang interpolation estimate of ∇u that we denote by $SZ(\nabla u)$ [234]. Indeed, since $\nabla u \in H^1(\Omega)$, there holds $\nabla u|_\Gamma \in H^{\frac{1}{2}}(\Gamma)$ and we are not able to assess it is continuous, in order to define its Lagrange interpolant. Then we bound the third term as follows, for $F \subset \Gamma$ a boundary facet with associated simplex T:

$$\|\nabla u - \nabla(\mathscr{I}^h u)\|_{0,F}$$
$$\le \|\nabla u - SZ(\nabla u)\|_{0,F} + \|SZ(\nabla u) - \nabla(\mathscr{I}^h u)\|_{0,F}$$
$$\le \|\nabla u - SZ(\nabla u)\|_{0,F} + c_i^{\frac{1}{2}} h_T^{-\frac{1}{2}} \|SZ(\nabla u) - \nabla(\mathscr{I}^h u)\|_{0,T}$$
$$\le \|\nabla u - SZ(\nabla u)\|_{0,F} + c_i^{\frac{1}{2}} h_T^{-\frac{1}{2}} \left(\|SZ(\nabla u) - \nabla u\|_{0,T} + \|\nabla(u - \mathscr{I}^h u)\|_{0,T} \right)$$
$$\le Ch_T^{\frac{1}{2}} \|\nabla u\|_{1,\omega_T} + h_T^{-\frac{1}{2}} \left(Ch_T \|\nabla u\|_{1,\omega_T} + Ch_T |u|_{2,T} \right)$$
$$\le Ch_T^{\frac{1}{2}} \|u\|_{2,\omega_T},$$

The notation ω_T stands for the patch of simplices in the neighbourhood of T, see [234]. We used first a triangular inequality, then (5.8), once again a triangular inequality, and finally approximation and interpolations errors, for the Scott-Zhang operator and the Lagrange operator, respectively. We sum on the boundary facets and this ends the proof. □

5.4   Implementation

Let us explicit how the extra terms in the Nitsche formulation (5.7) can be computed. We test Nitsche formulation (5.5) by taking, for every index $i \in 1, \dots, N$, the basis function φ_i as a test function ($v^h = \varphi_i \in V^h$). We get:

$$\sum_{j=1,\dots N} (k_{ij} - n_{ij} - \theta n_{ji} + \gamma_0 m_{ij}) \, u_j = f_i + \sum_{j=1,\dots N} (\gamma_N m_{ij} - \theta n_{ji}) g_j, \quad \forall i = 1, \dots, N.$$

Above we used the following notations for Nitsche matrices:

$$N_\Gamma = (n_{ij})_{i,j=1,\dots,N}, \quad n_{ij} = \int_\Gamma \varphi_i (\partial_\mathbf{n} \varphi_j),$$

and

$$M_\Gamma = (m_{ij})_{i,j=1,\dots,N}, \quad m_{ij} = \sum_{F \subset \Gamma} \int_F \frac{1}{h_F} \varphi_i \varphi_j.$$

Finally, Problem (5.7) recasted as a linear system is

$$\left(K - N_\Gamma - \theta N_\Gamma^\mathsf{T} + \gamma_N M_\Gamma \right) U = F + \left(\gamma_N M_\Gamma - \theta N_\Gamma^\mathsf{T} \right) G.$$

The discrete coercivity of Nitsche formulation ensures that the global Nitsche matrix is definite and positive, symmetric when $\theta = 1$. For low-order Lagrange finite elements, assembly of the linear system has no specific difficulties, and quadrature errors can be avoided. For the symmetric variant $\theta = 1$, a specific solver for symmetric matrices can be used, such as conjugate gradient or Cholesky method. For $\theta \neq 1$, any Krylov solver adapted for non-symmetric sparse systems can be used, or any variant of LU decomposition as implemented in many software.

5.5   Numerical Illustration

We consider the same test case as in previous chapter, and the numerical illustration is still done with scikit-fem [159]. The domain Ω is the unit square: $\Omega = (0, 1)^2$. The closed-form solution is still:

$$u(x, y) = \sin(\pi x) \sin(\pi y)$$

on Ω. Figure 5.2 depicts the solution obtained with Nitsche method, using the symmetric variant and $\gamma_0 = 1000$. The mesh and the finite element space are the same as in previous chapter. The global errors are approximately of 3.10^{-4} and

Fig. 5.2 Finite element
solution using the symmetric
Nitsche method

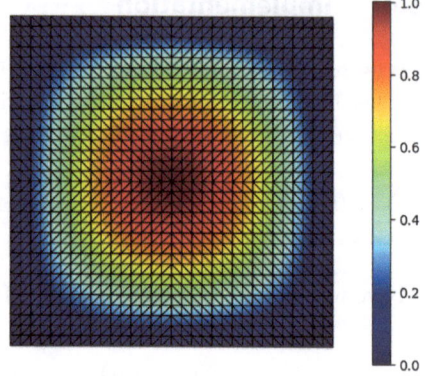

5.10^{-3}, in the L^2-norm and H^1-norm, respectively, and are comparable to the errors
with the standard method of the previous chapter. See Appendix A for details and
[78] for more tests.

5.6   Further Comments

For more details about Nitsche method, one can refer to the original paper of J.A.
Nitsche [209], to the monographs [86, 128, 242] or to the surveys [78, 163, 237].

5.6.1   Nitsche for Various Mathematical Models

For an abstract viewpoint with a methodology to derive Nitsche method for a large
class of, linear and non-linear, essential boundary or interface conditions and other
differential operators, see [170] (see also [86] for contact and friction conditions).
For a recent comparison in the case of interface problems with non-matching
meshes, see [52]. For an adaptation to fourth-order differential operator (Kirchhoff
plate), see [162]. For an adaptation to non-linear conditions defined on the domain
Ω, and not on the boundary, see [70, 160] for the obstacle problem and [83] for the
elastoplastic torsion problem.

For imposing various conditions in incompressible fluids, especially in the case
of curved boundaries, Nitsche method has proven its usefulness and has been
object of some recent papers. For slip conditions that are generalized Dirichlet
conditions for vector-valued functions, see, for instance, [15, 29, 151, 243]. For
Navier conditions, see [246], that is based on an extension of Nitsche for Robin
boundary conditions performed in [178]. For incompressible elasticity, see [36].

5.6.2 Nitsche Parameter and Condition Number

The value of γ_0 influences the condition number of the global stiffness matrix associated to $A_{N,\theta}(\cdot, \cdot)$, and for this reason it does not have to be taken too large. Anyway, since the method is consistent, the impact of the numerical parameter γ_0 has on the approximation of the Dirichlet condition is not as important as for the boundary penalty method. Moreover, since the function γ scales as $O(h^{-1})$ whatever the polynomial order k is, this does not deteriorate too much the conditioning that remains in $O(h^{-2})$ [68, 78, 163]. This property makes Nitsche method particularly relevant for high order discretizations, and for this reason, it has been considered for isogeometric analysis (IGA) [14, 77, 170], spline-based finite elements [123], or also for hybrid high-order (HHO) [73, 80] variational approximations.

5.6.3 Low Regularity and Other Estimates

The regularity assumption $u \in H^2(\Omega)$ in the a priori error estimate of this chapter is a technical assumption, to use the consistency in a straightforward manner. It is a delicate issue to weaken this assumption. However, and recently, optimal error estimates in the H^1-norm have been obtained even for solutions with low (minimal) regularity close to $H^1(\Omega)$. The interested reader can refer to, e.g. [126, 161]. They make use of techniques that are non-standard in the numerical analysis of finite element methods. Additionally, error estimates in the L^2-norm can be obtained; see e.g. [128, 242] for statements and proofs.

5.6.4 Nitsche Method as a Special Mixed Method

As pointed out by R. Stenberg in [237], Nitsche method is closely related to the stabilized mixed method of H. Barbosa and T.J.R. Hughes [30] (see also [34, 78, 86, 161, 177] for further considerations). The symmetric Nitsche method can be built alternatively from an augmented Lagrangian formalism; see [69, Section 5.2.2].

Problems

We consider once again the interface problem (Problem (2.34)) from Chap. 2 that we recall below:

$$\begin{cases} -\text{div}\,(k\nabla u) = f & \text{in } \Omega, \\ u = 0 & \text{on } \Gamma. \end{cases} \tag{5.13}$$

As in Chap. 2, we split the domain Ω into two subdomains Ω_1 and Ω_2, and we denote by $k_1 = k|_{\Omega_1}$ and $k_2 = k|_{\Omega_2}$, respectively, the value of the conductivity

Fig. 5.3 Two different
meshes for the interface
problem

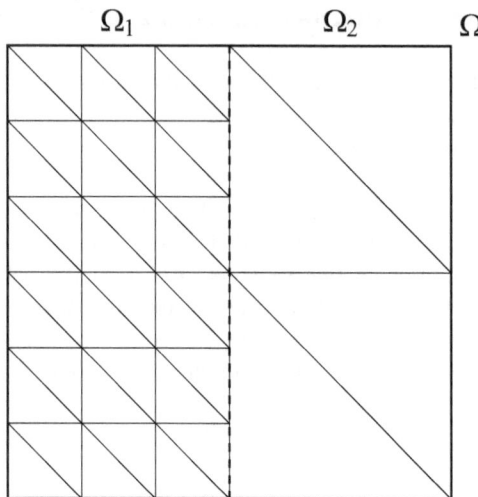

Ω_1 Ω_2 Ω

k on each subdomain The above problem can be reformulated by expliciting the
transmission conditions on the interface Σ between the two subdomains (see (2.35)
and (2.36) in Chap. 2):

$$\begin{cases} -\text{div } (k_i \nabla u_i) = f_i & \text{in } \Omega_i, \ i = 1, 2 \\ u_i = 0 & \text{on } \Gamma_i, \ i = 1, 2 \\ [\![u]\!] = 0 & \text{on } \Sigma, \\ \sigma_1 + \sigma_2 = 0 & \text{on } \Sigma. \end{cases} \tag{5.14}$$

Above the notation

$$[\![u]\!] := u_1 - u_2$$

stands for the jump of the solution on the interface Σ, and the notation $\sigma_i := k_i \nabla u_i \cdot n_i$, $i = 1, 2$, stands for the normal flux outwards to the boundary, respectively, to
the domain Ω_i.

We mesh the two subdomains with $\mathcal{T}_1^{h_1}$ and $\mathcal{T}_2^{h_2}$ of different sizes h_1 and h_2, that
do not match at the interface Σ. See Fig. 5.3. As a result, the whole mesh is not
an admissible mesh. The corresponding subspaces of finite elements are denoted by
V_1^h and V_2^h, and corresponding functions are noted $u_1^h, u_2^h, v_1^h, v_2^h$.

5.1 Following similar steps as for a Dirichlet boundary condition, suggest a Nitsche
method for the above problem.

5.2 Prove that the Nitsche method you obtain is consistent and well-posed provided
a condition on Nitsche parameter.

Nitsche for Signorini

6

This chapter presents a simple introduction to Signorini conditions for a membrane equation, as a prototype of non-linear non-smooth boundary conditions that model contact.

6.1 Outline

This chapter embraces almost every step of the pipeline below, Fig. 6.1, except postprocessing, since we start from a real-life system in which there is unilateral contact.

6.1.1 Contact Conditions

Indeed, instead of Dirichlet or Neumann boundary conditions, the following conditions on a solution u can be imposed on a portion of the boundary Γ of a domain Ω:

$$u \leq 0, \quad \partial_n u \leq 0, \quad u \partial_n u = 0. \tag{6.1}$$

These conditions can model the contact between a thin membrane and a rigid support. The unknown u here represents the vertical displacement of the thin membrane that is still subjected to the Laplace equation $-\Delta u = f$ on the whole domain Ω. Dirichlet or Neumann conditions can be applied on the other part of the boundary. The rigid support is supposed to occupy the region corresponding to $u \geq 0$ on the boundary. The condition $u \leq 0$ means that the membrane cannot penetrate into the obstacle. The condition $\partial_n u \leq 0$ means that the response of the obstacle is repulsive, and the complementarity condition $u \partial_n u = 0$ prevents unphysical situations in which there would be no effective contact $u < 0$, but still

© The Author(s), under exclusive license to Springer Nature Switzerland AG 2025
F. Chouly, *Finite Element Approximation of Boundary Value Problems*,
Compact Textbooks in Mathematics, https://doi.org/10.1007/978-3-031-72530-2_6

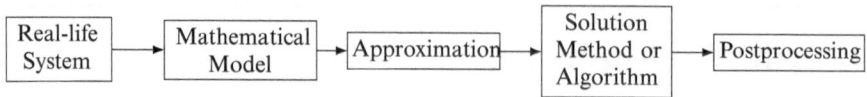

Fig. 6.1 We enter once again into the pipeline starting from another real-life system

a contact force $\partial_n u < 0$. These conditions are non-linear, so the whole boundary value problem is non-linear, even if there is a linear differential operator in the bulk.

6.1.2 A Variational Inequality

The set K of (kinematically) admissible displacements associated with contact conditions is made of the (regular enough) functions v that satisfy the non-penetration condition $v \leq 0$ on the contact boundary. As a result, K is not a vector space, but a convex cone. This is different from the set V_0 of Chap. 2 associated with a Dirichlet boundary condition that had the structure of a vector space. The membrane contact problem, also called scalar Signorini problem, in weak form, consists then in finding a function $u \in K$ solution to

$$\int_\Omega \nabla u \cdot \nabla (v - u) \geq \int_\Omega f(v - u), \tag{6.2}$$

for all $v \in K$. The above weak form is called a variational inequality. It generalizes the variational equation we have seen in Chap. 2: for an admissible set of solutions that is a vector space, we recover the variational equation. It can be shown [60, 86, 120, 152, 181] that this variational equation is well-posed, as a consequence of Stampacchia's Theorem: it admits one unique solution, and there is Lipschitz-continuity with respect to the data. Stampacchia's Theorem generalizes in fact the Lax-Milgram Theorem of Chap. 5, see [60].

Here the bilinear form associated with the Laplace operator is symmetric. So, still with the notation $\mathcal{J}$ for the energy functional associated with the membrane equation

$$\mathcal{J}(v) := \frac{1}{2} \int_\Omega \nabla v \cdot \nabla v - \int_\Omega f v,$$

the Stampacchia's Theorem ensures also in this situation that the unique solution u to the above variational inequality is as well the unique minimizer of the functional $\mathcal{J}$ on the convex set K. We recover in this situation the theorem of projection onto a convex set in a Hilbert space [60] or alternatively the existence and uniqueness of a global minimizer for a strongly convex function [9].

6.1.3   Standard Finite Element Approximation

The standard finite approximation for scalar Signorini consists in approximating directly the variational inequality (6.2). For this purpose, we build once again a discrete space V^h, with lowest order Lagrange finite elements. Then, we approximate the convex cone K of admissible displacements with a convex cone of finite dimension K^h. The most obvious choice is to set $K^h = V^h \cap K$, but there can be other possibilities. Then it suffices to find $u^h \in K^h$ that solves the discrete counterpart of (6.2).

Falk's Lemma, which generalizes the Cea's Lemma in the context of variational inequalities, allows to derive an a priori error bound. Here the contact terms make it difficult to obtain an optimal error estimate of $O(h)$ for a regular enough solution. With a simple argument, it is easy to reach $O(h^{\frac{3}{4}})$. Much more involved analysis is necessary to establish the optimal rate [86, 113].

6.1.4   Nitsche Approximation

Alternatively, we can reformulate contact conditions (6.1) as

$$\partial_n u = -[\gamma u - \partial_n u]_+, \tag{6.3}$$

for any positive function $\gamma > 0$ on the contact boundary and where $[x]_+ = \max(0, x)$ denotes the positive part of a scalar x. This can be used to formulate a Nitsche method, analogous to the method of Chap. 5, and that reads, in its simplest version

$$\begin{cases} \text{Find } u^h \in V^h \text{ such that:} \\ \displaystyle\int_\Omega \nabla u^h \cdot \nabla v^h + \int_{\Gamma_C} [\gamma u^h - \partial_n u^h]_+ v^h = \int_\Omega f v^h, \quad \forall\, v^h \in V^h, \end{cases}$$

where Γ_C is the contact boundary. Nitsche method transforms here a problem with inequality constraints into an unconstrained problem. The variational equation above is non-linear, but can be solved in practice using for instance a semi-smooth Newton method. The properties of the positive part allow to assess the well-posedness of the discrete formulation. Moreover, an optimal convergence rate of $O(h)$ can be derived with rather standard techniques; see [81, 84, 86].

6.2   The Scalar Signorini Problem

We present the scalar Signorini problem as a prototype of boundary value problem with a non-linear boundary condition. We detail the strong form and recast it in weak form as a variational inequality.

6.2.1   The Problem in Strong Form

Let Ω be a bounded open set in $\mathbb{R}^2$, of polygonal boundary $\partial\Omega$, that represents the shape of an elastic membrane. The boundary $\partial\Omega$ is divided into three different sets: Γ_D stands for the Dirichlet boundary, where the displacement on the boundary is prescribed, Γ_N stands for the Neumann boundary, where forces can be prescribed, and Γ_C is the boundary where contact can occur with a rigid support. Both the Dirichlet boundary Γ_D and the contact boundary Γ_N are supposed to have positive measure. The scalar Signorini problem reads

Find $u : \Omega \to \mathbb{R}$ solution to

$$-\Delta u = f \qquad \text{in } \Omega,$$
$$u = u_D \qquad \text{on } \Gamma_D, \qquad (6.4)$$
$$\partial_n u = h \qquad \text{on } \Gamma_N,$$

and with the following contact conditions on Γ_C:

$$u \leq 0, \quad (i)$$
$$\partial_n u \leq 0, \quad (ii) \qquad (6.5)$$
$$u\, \partial_n u = 0, \quad (iii).$$

with the notations $\Delta u = \text{div}\,(\nabla u)$, $\partial_n u := \nabla u \cdot n$ and with prescribed data $u_D \in H^{\frac{1}{2}}(\Gamma_D)$, $f \in L^2(\Omega)$, and $h \in L^2(\Gamma_N)$. This problem can be interpreted as follows: we find the vertical displacement u of a thin membrane that occupies the domain Ω, subjected to volumic forces f and boundary forces h, and that can come into contact with an obstacle on Γ_C. The obstacle occupies the position $u \geq 0$, so the condition (i) in (6.5) is a non-penetration condition: the membrane can be at distance from the obstacle ($u < 0$) or in contact with it ($u = 0$) but cannot penetrate it. The condition (ii) in (6.5) means that the contact force is repulsive, while the complementarity condition (iii) in (6.5) prevents states with non-penetration but non-zero contact forces. Remark that these conditions are analogous to Karush–Kuhn–Tucker conditions in constrained optimization [9, Chapter 10].

6.2.2   An Example of Closed-Form Solution

Before entering into the numerical approximation issues, let us first present a simple closed-form solution that comes from [73]. In the above problem (6.4)–(6.5), we take $f = 0$ (no bulk source term), $\Gamma_N = \emptyset$ (no Neumann boundary), and a rectangular domain

$$\Omega = (-1, 1) \times (-1, 0).$$

We define the contact boundary as the top part of the domain Ω

$$\Gamma_C = (-1, 1) \times \{0\}$$

and the remaining part of the boundary is Γ_D. For such a setting, we can report the closed-form solution of [73], which is, in polar coordinates:

$$u(r, \theta) = -r^\alpha \sin(\alpha\theta),$$

with $x = r\cos(\theta)$, $y = r\sin(\theta)$, $r \geq 0$, $0 \leq \theta \leq 2\pi$, and the value

$$\alpha = \frac{11}{2}.$$

On Γ_D, we set $u_D = u$, in order to satisfy the Dirichlet boundary condition. For this solution, a transition between biding and non-biding happens at point $(0, 0)$ on the contact boundary. This solution has been used to compute numerical convergence rates for a Nitsche hybrid high-order (HHO) method in [73].

6.2.3 Weak Formulation

The scalar Signorini problem admits a weak form as a variational inequality of the first kind and can be recast as a minimization problem [152].

Let us suppose, to simplify, that $h = 0$ and $u_D = 0$. Let us take u as a solution to (6.4)–(6.5) that is assumed regular enough. Let us start from (6.4) combined with the Green formula, and we get:

$$\int_\Omega \nabla u \cdot \nabla(v - u) = \int_\Omega f(v - u) + \int_{\Gamma_C} \partial_n u(v - u), \qquad (6.6)$$

where v is an arbitrary test function (virtual displacement). Denote by K the set of admissible displacements that satisfy the non-penetration condition i.e. for $v \in K$, there holds $v \leq 0$. We observe that K is a convex cone, and Signorini conditions (6.5) imply that $u \in K$. Now, for $v \in K$, still from Signorini conditions (6.5), there holds

$$\int_{\Gamma_C} \partial_n u(v - u) = \int_{\Gamma_C} \underbrace{\partial_n u}_{\leq 0} \underbrace{v}_{\leq 0} - \int_{\Gamma_C} \underbrace{(\partial_n u)u}_{=0} \geq 0.$$

This above inequality combined with the Green formula above yields

$$\int_\Omega \nabla u \cdot \nabla(v - u) \geq \int_\Omega f(v - u), \qquad (6.7)$$

for every test function (admissible virtual displacement) $v \in K$. The above weak
form (6.7) is a variational inequality, and this reformulation allows to prove that the
scalar Signorini problem has a sound mathematical structure. In addition, it can be
proven that every solution to the above variational inequality satisfies the problem
in the strong form [86, 181].

6.2.4 Well-Posedness Thanks to Stampacchia's Theorem

Before stating the well-posedness result, we recall Stampacchia's Theorem, which
is a cornerstone in the theory of variational inequalities. For its detailed proof, one
can refer to the original paper of J.-L. Lions and G. Stampacchia [190], or to [60,
Theorem 5.6].

Theorem 6.1 *Let V be a Hilbert space endowed with the inner product $(\cdot, \cdot)$ and
the norm $\| \cdot \| = \sqrt{(\cdot, \cdot)}$. Let K be a non-empty closed convex subset of V. Let $L(\cdot)$
be a continuous linear form on V. Let $a(\cdot, \cdot)$ be a continuous bilinear form on V.
Suppose that $a(\cdot, \cdot)$ is V-elliptic: there exists $\alpha > 0$ such that for any $v \in V$,*

$$a(v, v) \geq \alpha \|v\|^2.$$

Then, the variational inequality: find $u \in K$ such that

$$a(u, v - u) \geq L(v - u), \quad \forall v \in K,$$

*admits one unique solution. Moreover, if $a(\cdot, \cdot)$ is symmetric, then the solution u
to the above variational inequality is the unique minimizer on K of the quadratic
functional*

$$\mathcal{J} : K \ni v \mapsto \frac{1}{2} a(v, v) - L(v) \in \mathbb{R}.$$

Note first that we recover Lax-Milgram Theorem 5.1 in the particular case
where K is a vector space. In addition, if the bilinear form $a(\cdot, \cdot)$ is symmetric,
Stampacchia's Theorem is a reformulation of the result of existence and uniqueness
of the projection onto a closed convex set in a Hilbert space.

Let us recall the notations

$$a(v, w) = \int_{\Omega} \nabla v \cdot \nabla w, \quad L(w) = \int_{\Omega} f w,$$

for $v, w \in H^1(\Omega)$. We recall the notation $V = H^1(\Omega)$ and define now precisely the
convex set K of admissible displacements as

$$K := \{v \in V \mid v|_{\Gamma_D} = 0, \ v|_{\Gamma_C} \leq 0\}, \tag{6.8}$$

where $v|_{\Gamma_D}$ and $v|_{\Gamma_C}$ denote the trace of v on the Dirichlet and contact boundaries, respectively. The previous weak formulation for the scalar Signorini problem can be recasted as

Find $u \in K$ that satisfies

$$a(u, v - u) \geq L(v - u) \quad \text{for all } v \in K. \tag{6.9}$$

And now we can state the well-posedness of the scalar Signorini problem

Theorem 6.2 *Let us suppose that Ω is a connected bounded open set in $\mathbb{R}^2$, with boundaries Γ_D and Γ_C of positive measures. Let us suppose that $f \in L^2(\Omega)$, $u_D = 0$ and $h = 0$. Then, Problem (6.9) admits one unique solution $u \in K$. This solution is as well the unique minimizer of $\mathcal{J}(\cdot)$ on the convex cone K.*

Proof We apply Stampacchia's Theorem 6.1. An important issue is to prove that the non-empty convex set K is closed. This is technical and is done, for instance, in [86, 181, 183]. Also, the ellipticity of the bilinear form $a(\cdot, \cdot)$ is a consequence of the Deny-Lions Theorem 2.8. The other assumptions can be verified directly.   □

It can be also established that there is a Lipsichitz continuity dependence of the solution to the Signorini problem with respect to the data; see [86] for details.

6.3   A Discrete Variational Inequality

First, let us describe a direct or standard finite element approximation for the scalar Signorini problem.

6.3.1   Approximation with Finite Elements

We introduce $\mathcal{T}^h$ as a simplicial mesh of Ω, built as in Chap. 3, and that satisfies the shape-regularity assumption. We consider the lowest order Lagrange finite element space on this mesh, as in Chap. 3:

$$V^h := \left\{ v^h \in \mathscr{C}^0(\overline{\Omega}) \,\middle|\, v^h|_T \in \mathbb{P}_1(T), \forall T \in \mathcal{T}^h \right\}$$

We define the discrete convex cone as

$$K^h := K \cap V^h,$$

and introduce the discrete problem

$$\begin{cases} \text{Find } u^h \in K^h \text{ such that:} \\ a(u^h, v^h - u^h) \geq L(v^h - u^h), \quad \forall v^h \in K^h. \end{cases} \tag{6.10}$$

As a result, we discretize the variational inequality (6.9) directly, using a finite dimensional subset of the convex cone K of admissible displacements. There are many possibilities, in fact, to design some discrete convex cones; see [86, 113, 181].

6.3.2 A Priori Error Estimate

Stampacchia's Theorem 6.1 ensures that Problem (6.10) is well-posed as its continuous counterpart. Now let us try to study the approximation error. First, the following abstract error estimate is in the spirit of Falk's Lemma [131]:

Proposition 6.1 *Suppose that the solution* $u \in K$ *to Problem* (6.9) *belongs to* $H^2(\Omega)$. *Then the solution* u^h *to Problem* (6.10) *satisfies the* a priori *error estimate:*

$$\|u - u^h\|_{1,\Omega} \le C \inf_{v^h \in K^h} \left(\|u - v^h\|_{1,\Omega} + \left(\int_{\Gamma_C} \partial_n u (v^h - u) \right)^{\frac{1}{2}} \right), \qquad (6.11)$$

with $C > 0$ *independent from* h *and* u.

Proof Let $v^h \in K^h$. We use the ellipticity of $a(\cdot, \cdot)$, combined with Cauchy-Schwarz inequality:

$$\alpha \|u - u^h\|_{1,\Omega}^2 \le a(u - u^h, u - u^h)$$
$$= a(u - u^h, (u - v^h) + (v^h - u^h))$$
$$\le C \|u - u^h\|_{1,\Omega} \|u - v^h\|_{1,\Omega} + a(u - u^h, v^h - u^h),$$

where $\alpha > 0$ is the ellipticity constant of $a(\cdot, \cdot)$. Then we use the Young inequality

$$\frac{\alpha}{2} \|u - u^h\|_{1,\Omega}^2 \le \frac{C^2}{2\alpha} \|u - v^h\|_{1,\Omega}^2 + a(u, v^h - u^h) - a(u^h, v^h - u^h). \qquad (6.12)$$

Since v^h belongs to K^h, and u^h is the solution to (6.10):

$$-a(u^h, v^h - u^h) \le -L(v^h - u^h).$$

From the Green formula, we get

$$a(u, v^h - u^h) = L(v^h - u^h) + \int_{\Gamma_C} \partial_n u (v^h - u^h).$$

We combine the two previous relationships with the estimate (6.12):

$$\frac{\alpha}{2}\|u - u^h\|_{1,\Omega}^2 \le \frac{C^2}{2\alpha}\|u - v^h\|_{1,\Omega}^2 + \int_{\Gamma_C} \partial_n u(v^h - u^h).$$

There remains to transform the last term, as follows:

$$\int_{\Gamma_C} \partial_n u(v^h - u^h) = \int_{\Gamma_C} \partial_n u(v^h - u) + \int_{\Gamma_C} (\partial_n u)u - \int_{\Gamma_C} (\partial_n u)u^h.$$

We use the contact conditions (6.5) and the property $u^h \le 0$ (since $u^h \in K^h$)

$$\int_{\Gamma_C} (\partial_n u)u = 0, \quad -\int_{\Gamma_C} (\partial_n u)u^h \le 0.$$

We obtain finally

$$\frac{\alpha}{2}\|u - u^h\|_{1,\Omega}^2 \le \frac{C^2}{2\alpha}\|u - v^h\|_{1,\Omega}^2 + \int_{\Gamma_C} \partial_n u(v^h - u),$$

which gives the desired error bound (6.11).   □

Therefore, we can obtain genuinely a first estimation of the convergence rate for the finite element approximation (6.10). Indeed, we make the choice $v^h = \mathscr{I}^h u$, where $\mathscr{I}^h$ is the Lagrange interpolation operator introduced in Chap. 3. We verify $\mathscr{I}^h u \in K^h$ since $u \in K$ and since $\mathscr{I}^h$ preserves the positivity for $\mathbb{P}_1$ Lagrange finite elements. As a result, there holds:

$$\mathscr{I}^h(u) \le 0.$$

Now, let us bound the two right terms in the estimate (6.11). The first term corresponds to the interpolation estimate in the H^1-norm within the domain Ω and is classical. We just apply Theorem 3.1:

$$\|u - \mathscr{I}^h u\|_{1,\Omega} \le Ch|u|_{2,\Omega}. \tag{6.13}$$

For the second term of (6.11), which is the boundary term that comes from the contact condition on Γ_C, the simplest bound is obtained thanks to Cauchy-Schwarz inequality and then from an interpolation estimate on the trace of the solution [86, 115, 127]:

$$\int_{\Gamma_C} (\partial_n u)(\mathscr{I}^h(u) - u) \le \|\partial_n u\|_{0,\Gamma_C} \|\mathscr{I}^h(u) - u\|_{0,\Gamma_C}$$

$$\le C\|\partial_n u\|_{0,\Gamma_C} h^{\frac{3}{2}}\|u\|_{\frac{3}{2},\Gamma_C}$$

where we used the property (3.9). We take the square root and get

$$\left(\int_{\Gamma_C} \partial_n u (\mathscr{I}^h u - u) \right)^{\frac{1}{2}} \leq Ch^{\frac{3}{4}} \|\partial_n u\|_{0,\Gamma_C}^{\frac{1}{2}} \|u\|_{\frac{3}{2},\Gamma_C}^{\frac{1}{2}}.$$

Finally, we combine the above result with (6.13) and obtain

$$\|u - u^h\|_{1,\Omega} \leq Ch^{\frac{3}{4}} \|u\|_{2,\Omega}. \tag{6.14}$$

The above result is suboptimal, since it provides only a convergence rate $O(h^{\frac{3}{4}})$, instead of the expected rate $O(h)$ that comes from the interpolation estimate and that we find usually for elliptic problems with standard boundary conditions; see Chap. 4.

This is the starting point of a long story that started in the year 1970 with, at first, such suboptimal bounds. Much effort has been dedicated since then to improve these first results. Finally, the first optimal result has been proven in 2015; see [113]. The interested reader can refer, for instance, to [152] for error bounds on various variational inequalities, and to the bibliography in [86, 113] for more details about the intermediate results between the first estimations and the optimal ones.

6.4 Nitsche for Signorini

The motivation to find some alternative discretizations for the Signorini problem is twofold: to get methods attractive for implementation and to have better mathematical properties of stability, robustness, and convergence. As a result, we present here an extension of the Nitsche method presented in Chap. 5. More details can be found in [86] and references therein.

6.4.1 Building a Nitsche Method for Signorini

The first observation to derive a Nitsche method for the scalar Signorini problem (6.4)–(6.5) is the following: we can reformulate the contact condition as a nonlinear inequality. Indeed, for any positive function $\gamma > 0$ on the contact boundary, the Signorini conditions (6.5) are equivalent to

$$\partial_n u = -[\gamma u - \partial_n u]_+. \tag{6.15}$$

This is a direct consequence of the reformulations of Kuhn-Tucker conditions $a \leq 0$, $b \leq 0$, $ab = 0$ as $0 = \max(a, b)$, for $a, b \in \mathbb{R}$ (see [86] for a detailed proof).

Now we start again from (6.4) with $u_D = 0$ and $h = 0$, and we apply the Green formula

$$\int_\Omega \nabla u \cdot \nabla v - \int_{\Gamma_C} \partial_n u v = \int_\Omega f v,$$

for a test function v. It remains to inject the condition (6.15)

$$\int_\Omega \nabla u \cdot \nabla v + \int_{\Gamma_C} [\gamma u - \partial_n u]_+ v = \int_\Omega f v$$

and we get the basic formulation for an incomplete Nitsche method. After discretization, we get the following Nitsche weak form:

$$\begin{cases} \text{Find } u^h \in V^h \text{ such that:} \\ \int_\Omega \nabla u^h \cdot \nabla v^h + \int_{\Gamma_C} [\gamma u^h - \partial_n u^h]_+ v^h = \int_\Omega f v^h, \quad \forall v^h \in V^h. \end{cases}$$

The function $\gamma > 0$ is chosen in the same way as in the previous chapter. For consistency, well-posedness, and optimal error bound associated with this method, the interested reader may refer to [86] and references therein.

6.4.2 Solution with a Semi-smooth Newton Method

There remains to treat the non-linearity for the above Nitsche formulation for Signorini contact. A popular and efficient way is to make use of a semi-smooth Newton technique. Let us first define the residual

$$\mathcal{R}(u^h; v^h) := \int_\Omega f v^h - \int_\Omega \nabla u^h \cdot \nabla v^h - \int_{\Gamma_C} [\gamma u^h - \partial_n u^h]_+ v^h,$$

for $v^h \in V^h$. To derive the tangent system, we use the relationship

$$\frac{\mathrm{d}}{\mathrm{d}x}[x]_+ = H(x),$$

where $x \in \mathbb{R}$ and $H(\cdot)$ is the Heaviside function. As a result, at each Newton iteration, we need to

$$\begin{cases} \text{Find } \delta u^h \in V^h \text{ such that:} \\ \int_\Omega \nabla \delta u^h \cdot \nabla v^h + \int_{\Gamma_C} H(\gamma u^h - \partial_n u^h)(\gamma \delta u^h - \partial_n \delta u^h) v^h \\ = -\mathcal{R}(u^h; v^h) \quad \forall v^h \in V^h. \end{cases}$$

There remains then to carry out the update $u^h + \delta u^h \to u^h$ and to move to the next iteration.

6.5   Further Comments

For more details about variational inequalities and contact problems, the interested reader can refer, for instance, to the monographs or review papers [81, 86, 120, 121, 152, 164, 181, 183, 235, 248].

6.5.1   More Mathematical Models of Contact

The scalar Signorini problem is almost the simplest model for unilateral contact and is described for instance in [152]. More realistic models may involve (small strain or large strain) elasticity, in which the partial differential equation involve a vectorial unknown. In addition, instead of contact between an elastic body and a rigid support, contact between various elastic bodies can be modelled. Last but not least, contact can be complemented by a friction law. The most common friction laws are Tresca friction and Coulomb's friction. All these models are detailed in e.g. [86, 120, 181]. Particularly, in [181], various models of friction are described. For recent results about Nitsche method applied to contact and friction problems, see e.g. [81, 85, 86].

6.5.2   Numerical Approximation Techniques

There are many more possibilities to approximate contact conditions than the two ones described in this chapter. Let us mention notably:

- Penalization or regularization of contact conditions, as described and studied, for instance, in [181]; see also [182, 211]
- Mixed and mortar techniques and variants, based on the appropriate design of dual spaces of Lagrange multipliers; see, for instance, [1, 43, 86, 164, 169, 248]

For semi-smooth Newton, or generalized Newton applied to contact, especially discretized with finite elements and Nitsche, see, for instance, [7, 86, 223].

A Posteriori Error Estimation

7

This chapter is about practical situations where the user is interested to estimate the error due to the finite element approximation.

7.1 Outline

At the end of the pipeline, there is the postprocessing of the solution delivered by the computer (last but not least); see Fig. 7.1.

Maybe you have realized that this pipeline is mostly based on a logical and mathematical structure and does not match to the pipeline when one uses a commercial finite element software (which can be thought differently for contingent and practical reasons). But at least, it may allow to understand what is behind the blackbox.

Mathematical properties such as well-posedness or optimal a priori error bounds are important properties of a *method* that ensure it has no pathology. Nevertheless, they are sometimes not enough for real practitioners.

7.1.1 Error Control Through A Posteriori Error Estimation

This chapter presents the idea of a posteriori error estimators that emerged at the end of the 1970s and has been an object of intense research activity since the end of the 1990s.

Since a practitioners carry out a simulation on a given mesh of size h, and since this simulation takes a given amount of time, it can be of interest for them to evaluate that, with the computational resources they have access to, they are able to compute an approximate solution with the desired accuracy, i.e. to know if the approximation error $(u - u^h)$ in a given norm is below a given tolerance ε. Naively, they cannot evaluate the approximation error directly since u is unknown. However, various

© The Author(s), under exclusive license to Springer Nature Switzerland AG 2025   111
F. Chouly, *Finite Element Approximation of Boundary Value Problems*,
Compact Textbooks in Mathematics, https://doi.org/10.1007/978-3-031-72530-2_7

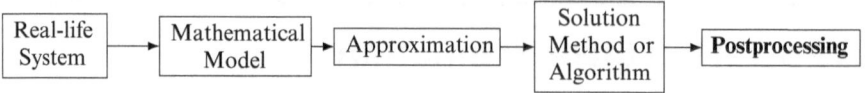

Fig. 7.1 The above pipeline depicts the global process behind a numerical simulation. We are ending with postprocessing

techniques have been found in order to deliver a quantity $\eta(=\eta(u^h))$ that provides an estimation of the discretization error and that can be computed using only the discrete solution u^h. This quantity η is called an a posteriori error estimator.

Moreover, the basic characteristics of the finite element method allows to split easily the global quantity η into a sum of local quantities η_T associated with each simplex T in the mesh. These quantities allow to quantify the local contribution to the global error. These local contributions η_T allow to modify locally the mesh to minimize this error optimally. Roughly speaking, the simplices where η_T is considered too large are split into smaller simplices, and also patches of simplices associated with low values η_T can be glued to coarsen locally the mesh, in regions that do not need high accuracy.

7.1.2 Uniform vs. Adaptive Mesh Refinement

Indeed, we have seen in Chap. 2 that, for nonconvex domains, singularities appear at re-entrant corners and prevent the solution from being very regular. This implies a loss of efficiency when uniform refinement is performed (remember the assumptions in the previous chapters to get optimal theoretical convergence rates). This is also the case when the source term is singular: see, for instance, the one-dimensional example with a Dirac source term in [9, Chapter 6], where numerical convergence rates are in $O(h^{\frac{1}{2}})$, much below the optimal rate of $O(h)$.

Moreover, for boundary value problems closer to real life applications, other singularities, or let us say, at least, strong local variations of the solution, can appear for many reasons: boundary layers, singular boundary terms, singular coefficients of the partial differential equation, transitions between different boundary conditions, etc. In any of these situations, adaptive mesh refinement allows to modify the mesh only for spatial regions where accuracy is really needed.

It is clear that nowadays adaptive finite element methods have become rather standard within the theoretical numerical analysis community, at least from the beginning of 2000s. However, it is still less considered by many practitioners.

7.1.3 Goal-Oriented Error Estimations

In many applications, the global solution itself is not of interest, but only an output quantity J determined from the solution, computed, for instance, using the solution or its derivatives in a small subdomain. Examples can be the drag coefficient in fluid dynamics simulations or stress intensity factors in fracture mechanics. The idea of goal-oriented error estimation is to drive the mesh adaptation to minimize directly the error on the quantity of interest J, instead of targeting the global norm of the solution.

Roughly speaking, goal-oriented error estimators are generally based on the computation of an auxilliary problem (adjoint or dual problem) where the quantity J is considered a source term. This allows to compute an auxilliary function z over the domain that quantifies how sensitive J is to specific localized values of the discretization error. For each simplex T in the mesh, this quantity z allows to compute a weight, or local sensitivity factor, that we denote by $\omega_T(z)$, which allows to determine how much J is influenced by the approximation error on T. Then, mesh refinement is driven by local quantities of the form

$$\omega_T(z)\eta_T(u^h),$$

instead of $\eta_T(u^h)$. As a result, if $\omega_T(z)$ has a small value for a simplex T, mesh refinement will not be applied for the simplex T since the local approximation error has a very small impact on the specific quantity J.

7.2   Our Model Problem and Its Finite Element Approximation

To present basic concepts of a posteriori error estimation and adaptive mesh refinement, it is better to start from a well-defined and simple example. So we go back to Poisson's problem with Dirichlet boundary condition. We take a domain Ω that is an open bounded polytope of $\mathbb{R}^d$ ($d = 1, 2, 3$), with Lipschitz boundary $\Gamma := \partial\Omega$. We reintroduce the Problem (2.8) from Chap. 2 with some minor simplifications:

Find $u : \Omega \to \mathbb{R}$ that satisfies :

$$\begin{cases} -\Delta u = f & \text{in } \Omega, \quad (i) \\ \quad u = 0 & \text{on } \Gamma. \quad (ii) \end{cases} \qquad (7.1)$$

In comparison with Chap. 2, we took $\mu = 1$ and $g = 0$. We suppose that the source term f belongs to $L^2(\Omega)$. We recall that the bilinear form and linear form associated to (7.1) read

$$a(v, w) = \int_\Omega \nabla v \cdot \nabla w, \quad L(w) = \int_\Omega fw,$$

for $v, w \in H^1(\Omega)$. We recall the notation $V_0 = H_0^1(\Omega)$ for the functions in $H^1(\Omega)$ with vanishing trace on the boundary. The above problem in weak formulation reads

Find $u \in V_0$ that satisfies

$$a(u, v) = L(v) \quad \text{for all } v \in V_0. \tag{7.2}$$

We suppose that $\mathcal{T}^h$ is a simplicial mesh of Ω, built as in Chap. 3, and that satisfies the shape-regularity assumption. We reintroduced the lowest order Lagrange finite element space on this mesh, as in Chap. 3:

$$V^h := \left\{ v^h \in \mathscr{C}^0(\overline{\Omega}) \;\middle|\; v^h|_T \in \mathbb{P}_1(T), \forall T \in \mathcal{T}^h \right\}$$

and $V_0^h = V^h \cap V_0$. The standard finite element method is the same as in Chap. 4:

Find $u^h \in V_0^h$ that satisfies

$$a(u^h, v^h) = L(v^h) \quad \text{for all } v^h \in V_0^h. \tag{7.3}$$

Remember that both the continuous and discrete Problems (7.1) and (7.3) are well-posed, and provided that the solution u is regular enough ($u \in H^2(\Omega)$), we have the optimal a priori error bound (4.8) that we recall below:

$$\|u - u^h\|_{1,\Omega} \leq Ch\|u\|_{2,\Omega}, \tag{7.4}$$

However, for a nonconvex domain, for instance, we cannot expect the solution u to belong to $H^2(\Omega)$, because of the singularities that appear at re-entrant corners. The solution lies in an intermediate space between $H^1(\Omega)$ and $H^2(\Omega)$, and this induces a convergence rate in $O(h^\alpha)$ with $0 \leq \alpha \leq 1$.

More intuitively, it is clear that if the solution u to a boundary value problem such as (7.1) is not expected to be very smooth, uniform refinement is not the best technique to reduce the discretization error $\|u - u^h\|_{1,\Omega}$. Indeed, it tends to add unnecessary degrees of freedom in regions where the discrete solution is already very close to the exact one, and not enough degrees of freedom where the solution would need to be more resolved.

The idea of adaptive mesh refinement is to decrease locally the size of the simplices only in the regions where more accuracy is really needed. Also, the idea of a posteriori error estimation is to use the solution u^h to compute local quantities η_T on each simplex. These quantities will point out where accuracy is lacking. This allows notably to take fully advantage of the finite element technology that allows much flexibility related to the mesh, in comparison to some other discretization methods that may require much more geometrical restrictions, at least in their basic version.

By the way, if u is very smooth, one may consider higher-order methods instead of piecewise linear Lagrange finite elements to increase the accuracy without increasing too much the size of the linear system.

7.3 An Explicit Residual A Posteriori Error Estimator

We provide first the basic idea of a posteriori error estimation, and then detail the expression of a common specimen among them. For a simple introduction to the topic, see, for instance, [35, 63, 214] and classical textbooks [6, 210, 244].

7.3.1 Discrete Error and Residual

Let us consider a weak form such as (7.2). Since the bilinear form $a(\cdot, \cdot)$ is symmetric and elliptic, it defines an inner product on V_0. As a result, the discrete error or approximation error can be measured in the norm

$$\left(a(u - u^h, u - u^h) \right)^{\frac{1}{2}}.$$

Then, thanks to the Riesz-Fréchet Representation Theorem, the vector $u - u^h$ in V_0 associated with the approximation error is associated with a unique continuous linear form on V_0, and the above norm is equal to the dual norm

$$\left(a(u - u^h, u - u^h) \right)^{\frac{1}{2}} = \sup_{v \in V_0} \frac{a(u - u^h, v)}{(a(v, v))^{\frac{1}{2}}}. \tag{7.5}$$

Since u solves (7.2), the numerator of the above expression can be reformulated as

$$a(u - u^h, v) = a(u, v) - a(u^h, v) = L(v) - a(u^h, v) \tag{7.6}$$

and at this point, we obtain a quantity that does not depend on u. We call it the residual of the equation and define it as

$$\mathcal{R}(v) := L(v) - a(u^h, v). \tag{7.7}$$

With the following notation for the dual norm

$$\|\mathcal{R}\|_{\mathscr{L}(V_0, \mathbb{R})} := \sup_{v \in V_0} \frac{\mathcal{R}(v)}{(a(v, v))^{\frac{1}{2}}},$$

we combine the previous relationships (7.5)–(7.6)–(7.7) and get the following result:

$$\left(a(u - u^h, u - u^h) \right)^{\frac{1}{2}} = \|\mathcal{R}\|_{\mathscr{L}(V_0, \mathbb{R})}. \tag{7.8}$$

This means that $\|\mathcal{R}\|_{\mathscr{L}(V_0, \mathbb{R})}$ is (almost) the perfect candidate for a posteriori error estimation: it measures the discrete error only using the value of u^h and the data of the continuous boundary value problem. There remains two practical issues now:

1) finding a quantity close enough to $\|\mathscr{R}\|_{\mathscr{L}(V_0,\mathbb{R})}$ but more friendly for a computer and 2) localizing the contributions on the different simplices of the mesh, in order to do mesh refinement driven by minimization of the discrete error. Before addressing these issues, let us make a few observations:

1. The residual $\mathscr{R}$ is a linear form, on the continuous space V_0.
2. The residual $\mathscr{R}$ vanishes on the whole discrete subspace V_0^h because of Galerkin orthogonality.
3. If $\mathscr{R} = 0$, necessarily $u^h = u$ (and this means that $u \in V_0^h$). Conversely, if $u^h = u$, then $\mathscr{R} = 0$.
4. The smaller the residual is, the closer u^h is to u, and conversely.
5. Still because of Galerkin orthogonality, we have in fact that $\mathscr{R}(v) = \mathscr{R}(v - v^h)$ with v^h a given discrete approximation to v.

Finally, as an illustration, for Poisson's problem with Dirichlet boundary conditions (7.2), there holds:

$$\|\nabla(u - u^h)\|_{0,\Omega} = \sup_{v \in V_0} \frac{\int_\Omega \nabla(u - u^h) \cdot \nabla v}{\|\nabla v\|_{0,\Omega}}$$

and the expression (7.8) reads in this case

$$\|\nabla(u - u^h)\|_{0,\Omega} = \sup_{v \in V_0} \frac{\int_\Omega fv - \int_\Omega \nabla u^h \cdot \nabla v}{\|\nabla v\|_{0,\Omega}}.$$

In this particular case, the residual that is expressed as

$$\mathscr{R}(v) = \int_\Omega fv - \int_\Omega \nabla u^h \cdot \nabla v. \tag{7.9}$$

7.3.2 Definition of the A Posteriori Error Estimator

Let us describe an explicit residual-based *a posteriori* error estimate for Problem (7.3) and based on the previous considerations. We introduce standard notations:

- We define $\mathcal{F}_h$ the set of facets of the simplicial mesh and define

$$\mathcal{F}_h^{\text{int}} := \{F \in \mathcal{F}_h : F \subset \Omega\}$$

  as the set of interior facets of $\mathcal{T}^h$.

- For a simplex T, we denote by $\mathcal{F}_T$ the set of facets of T and according to the above notation, we set $\mathcal{F}_T^{int} := \mathcal{F}_T \cap \mathcal{F}_h^{int}$.
- For a facet F of a simplex T, introduce $n_{T,F}$ the unit outwards normal vector to T along F. Furthermore, for each facet F, we fix one of the two normal vectors and denote it by n_F.
- The jump of some function v across a facet $F \in \mathcal{F}_T^{int}$ at a point $x \in F$ is defined as

$$\llbracket v(x) \rrbracket_F := \lim_{\alpha \to 0^+} v(x + \alpha n_F) - v(x - \alpha n_F).$$

- Let ω_T be the union of all simplices having a nonempty intersection with T.
- f_T is a computable quantity that approximates f on the simplex $T \in \mathcal{T}^h$.

The *a posteriori* error estimator is defined as follows:

Definition 7.1 For Problem (7.3), the local error estimators η_T and the the global estimator η are defined by

$$\eta_T := \left(\eta_{1,T}^2 + \eta_{2,T}^2 \right)^{\frac{1}{2}},$$

$$\eta_{1,T} := h_T \| f_T \|_{0,T},$$

$$\eta_{2,T} := h_T^{\frac{1}{2}} \left(\sum_{F \in \mathcal{F}_T^{int}} \llbracket \nabla u^h \cdot n_F \rrbracket_F \right)^{\frac{1}{2}},$$

$$\eta := \left(\sum_{T \in \mathcal{T}^h} \eta_T^2 \right)^{\frac{1}{2}},$$

and the local and global approximation terms are given by

$$\zeta_T := \left(h_T^2 \sum_{T' \subset \omega_T} \| f - f_{T'} \|_{0,T'}^2 \right)^{\frac{1}{2}},$$

$$\zeta := \left(\sum_{T \in \mathcal{T}^h} \zeta_T^2 \right)^{\frac{1}{2}}.$$

The above expression is, in fact, derived from the fundamental relationship (7.9), where the integral terms in the residual are split into local terms on each simplex T, and where the Green formula is applied locally. See, for instance, [6, 244] for the detailed derivation.

The idea behind this definition is the following: the estimator η_T allows to estimate the contribution to the global error for a given simplex T. Roughly speaking, a mesh refinement method will split the simplices for which η_T is the largest, to refine locally.

Remark the jump term in the estimator. It is somehow linked to the idea that the gradient of the continuous solution has no jumps at the interfaces between the simplices, while the discrete solution presents these weak discontinuities for Lagrange finite elements.

The approximation terms allow to replace the source term f by an approximation term f_T easy to compute, for instance, a piecewise constant function on each simplex.

Finally, the global term η collects all the contributions from all the simplices and allows to provide an indication for the overall magnitude of the discretization error. This allows, for instance, to introduce a simple stopping criterion in iterative refinement procedures.

Remark 7.1 For the Nitsche method presented in Chap. 5, the above estimator can be modified in order to incorporate Nitsche extra terms, see e.g. [38]. Also, for Nitsche method extended to Signorini contact or to frictional contact (see Chap. 6), residual estimators are studied in [17, 82], and equilibrated fluxes estimators are designed in [110, 136].

7.4   Adaptive Finite Element Methods

The idea of adaptive finite element methods is to solve iteratively a boundary value problem such as (7.2). A first finite element approximation is computed on a very coarse mesh, and then, an a posteriori error estimator serves to detect the most important local contributions to the discretization error. The mesh is refined accordingly to the estimator and a new solution can be computed. The typical form of an adaptive finite element method is then:

$$\textbf{Solve} \quad \rightarrow \quad \textbf{Estimate} \quad \rightarrow \quad \textbf{Mark} \quad \rightarrow \quad \textbf{Adapt} \quad \rightarrow \quad \cdots$$

In the above diagram:

1. **Solve** consists in solving a problem such as Problem (7.3), using the mesh and the corresponding finite element space at hand.
2. **Estimate** consists in computing the value of the a posteriori error estimate η from its local contributions on simplices η_T. In general, if η is below a prescribed threshold, the solution is considered accurate enough and the adaptive algorithm is stopped.
3. **Mark** consists in labelling the simplices that are candidates for refinement and that correspond to local estimators η_T with the largest values. There are many possibilities to proceed within this step, and a common procedure consists in

marking a prescribed fraction of the simplices T that correspond to the largest values η_T.

4. **Adapt** consists in modifying the mesh in order to refine where the a posteriori error is the largest. For instance, the marked simplices can be split into smaller ones, and neighbours can be modified so that the mesh remains admissible. Once the mesh is refined, a new finite element space needs to be built and we can loop to the first step **Solve**.

There is an abundant literature on this topic, as there are many ways to proceed. See e.g. [35, 63, 210] and references therein for more details.

7.5   Upper and Lower Bounds

Two important mathematical properties of error estimators are reliability and efficiency. First, an upper bound can be given for the estimator of Definition 7.1; see e.g. [244].

Theorem 7.1 *Let $u \in V_0$ be the solution to Problem (7.2) and let u^h be the finite element solution of Problem (7.3). There holds*

$$\|u - u^h\|_{1,\Omega} \le C(\eta + \zeta),$$

where the constant $C > 0$ does not depend of the mesh size h.

The above result ensures that the estimator η is reliable, in the sense that if it decreases, then the real discrete error also needs to decrease. This property makes the estimator a good candidate to drive mesh refinement.

We now consider the local lower error bounds of the discretization error terms. For the proof, see also, for instance, [244].

Theorem 7.2 *For all the mesh simplices $T \in \mathcal{T}^h$, the following local lower error bounds hold:*

$$\eta_{1,T} \le C(\|u - u^h\|_{1,T} + \zeta_T), \tag{7.10}$$

$$\eta_{2,T} \le C(\|u - u^h\|_{1,\omega_T} + \zeta_T), \tag{7.11}$$

where $C > 0$ is a constant independent of the mesh size h.

The above result ensures that the estimator η is efficient: it cannot overestimate too much the local discretization error.

For a simple alternative presentation about these aspects, see particularly [214], as well as [6, 244] for the detailed proofs.

7.6   Goal-Oriented Error Estimation

Let us now do a small incursion into the world of goal-oriented error estimates. The idea behind this methodology is to control the error on a given quantity:

$$J : V_0 \ni v \to J(v) \in \mathbb{R},$$

instead of the natural (Sobolev) global norm of the solution. Indeed, a practitioner does a finite element computation to predict some user-defined output quantities and should be more concerned by the prediction of these quantities than the error for the global solution in the natural norm.

The basic idea of goal-oriented error estimation is to compute the solution to a dual problem. This solution will allow to know the regions that impact the most the accuracy for the quantity J and will be used to weight a standard a posteriori error estimator.

7.6.1   The Becker and Rannacher Technology

We present in a simplified manner the general framework in [39] called dual-weighted residuals (DWR). For linear boundary value problems and linear functionals, indeed, it can be outrageously simplified, and we follow the presentation in [116,228]. We suppose that $J \in L^2(\Omega)$ and introduce the dual problem: find $z \in V_0$ solution to

$$a(w, z) = J(w), \quad w \in V_0. \tag{7.12}$$

Then, to estimate the error on J, we proceed as follows:

$$J(u) - J(u^h) = a(u, z) - a(u^h, z) = L(z) - a(u^h, z)$$

where we used first (7.12) and then (7.1). So we get

$$J(u) - J(u^h) = \mathscr{R}(z) \tag{7.13}$$

where the residual $\mathscr{R}(z)(= L(z) - a(u^h, z))$ has been introduced previously in (7.7) and is (almost) a fully computable quantity. To end the process, there just remains:

1. To replace the exact dual solution z by an approximate computed solution
2. To apply localization techniques to split the dual residual into contributions associated with each simplex T in the mesh

More details about these steps can be found, for instance, in [39, 228] and particularly in [124]. Numerical illustrations for the computation of the dual solution can be found in [116].

7.6.2 The Becker, Estecahandy, and Trujillo Technology

Alternatively, as in [37], we can proceed as follows: We introduce a discrete dual problem using the same mesh and finite element space as the primal problem: find $z^h \in V_0^h$ solution to

$$a(w^h, z^h) = J(w^h), \quad \forall w^h \in V_0^h. \tag{7.14}$$

Note that, at this point, the discrete solution to this dual problem, z^h, cannot be used in the previous error estimator to approximate z, since the Galerkin orthogonality induces $\mathscr{R}(z^h) = 0$. This can be cumbersome in practice and one would like to use the same mesh, finite element space, and stiffness matrix to compute both the primal and the dual solutions. For this purpose, we can follow the next steps and first estimate, as previously

$$J(u) - J(u^h) = a(u, z) - a(u^h, z) = a(u - u^h, z).$$

where we used the (continuous) dual problem (7.12). Then, the Galerkin orthogonality for the primal problem reads

$$a(u - u^h, z^h) = 0.$$

As a result we get

$$J(u) - J(u^h) = a(u - u^h, z) - a(u - u^h, z^h) = a(u - u^h, z - z^h).$$

We finally use the Cauchy-Schwarz inequality and obtain

$$J(u) - J(u^h) \leq \|u - u^h\|_{1,\Omega} \|z - z^h\|_{1,\Omega}.$$

This leads this time to the bound

$$|J(u) - J(u^h)| \leq C \eta_u \eta_z + \varepsilon \tag{7.15}$$

where η_u is an a posteriori error estimator for the primal problem, and η_z is another error estimator for the dual problem. The notation ε stands for higher-order terms that come from the approximation terms. Both estimators are supposed to satisfy an upper bound property as in Theorem 7.1. Both can be, for instance, explicit residual error estimators described previously, as in [37]. Alternatively, they can be built from other error estimators: see, for instance, [64] where hierarchical error estimators are used instead of residual ones.

Fig. 7.2 Initial mesh for the
L-shaped domain

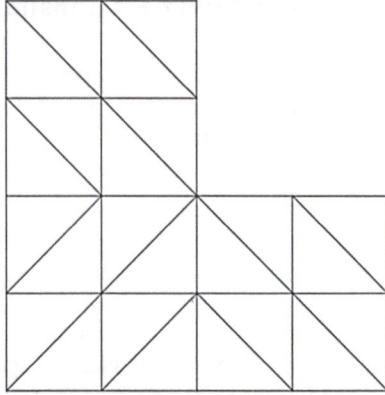

Fig. 7.3 Refined mesh and
final computed solution

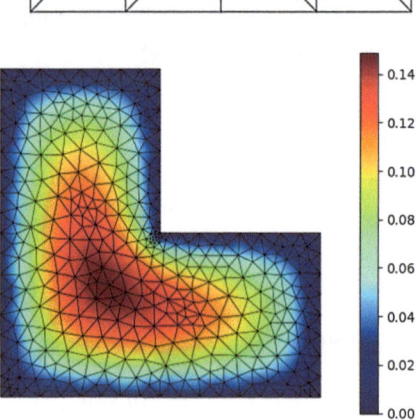

7.7   A Numerical Illustration

We end this chapter with a small numerical illustration for an L-shaped domain, made with scikit-fem and inspired from an example of the documentation.[1] Figure 7.2 depicts the initial mesh, while Fig. 7.3 depicts the final solution and the refined mesh after a few refinement iterations.

7.8   Further Comments

There are many recent works devoted to postprocessing and error control. Below we mention a few of them:

---

[1] See https://github.com/kinnala/scikit-fem/blob/9.1.1/docs/examples/ex22.py

7.8.1   Model Error

When a numerical simulation is carried out, the result can be different from what is observed in the real-life system. One simple example is weather forecast. The weather can be predicted by solving the equations that govern the ocean and the atmosphere (fluid dynamics equations, thermics, etc.) but even a supercalculator is not capable of predicting the weather with reliability for more than a few days.

In fact, each step of the pipeline (2.1) corresponds to opportunities to introduce some new errors that can pollute the solution delivered by the computer, sometimes to a point that it cannot be trusted or useful. The approximation error or discretization error takes into account the approximation of a continuous model by a discrete one (it measures the distance between u and u^h). But of course, the model may take into account imperfectly the intrinsic properties of the real-life system of interest.

Modelling errors are not easy to define if a direct comparison between the real-life system and the mathematical model is made. However, for instance, in classical mechanics/continuum mechanics, it is usual to have at least a hierarchy of models of increasing complexity. For instance, in fluid mechanics, from the non-stationary Navier-Stokes equations in three-dimensional geometries, simpler models can be derived: Prandtl/boundary layer equations, Oseen equations, Stokes equations, etc. Also, for some configurations, a three-dimensional problem can be reduced to a problem in dimension one or two. The same occurs in solid mechanics where various simplified models can be derived from three-dimensional large-strain hyperelasticity. A posteriori error estimations that take into account model errors have been proposed, for instance, in [56, 213, 227].

7.8.2   Numerical Errors

Numerical errors arise in the penultimate step of the pipeline, when some code is implemented in a computer to determine the solution of a finite element problem. The solution $\widetilde{u}^h$ delivered by the computer differs from the (theoretical) solution u^h to a finite element problem such as (7.3). This is mostly because of possible numerical integration errors, inexact solving of the linear system and roundoff errors. In some situations, the numerical errors are of small magnitude comparatively to the discretization error. In some other situations, the numerical errors can prevent the simulation to be reliable and accurate. See [214] and references therein for a discussion and the possibilities to take into account the numerical error into a posteriori error estimation.

7.8.3   More About Meshes

First, let us mention that anisotropic meshes, with simplices or cells that can be very elongated, are very useful for some classes of boundary value problems; see, for instance, [138, 139, 218].

In [117], the authors study the effect of having a few degenerate simplices in a mesh. Using a modified interpolation operator, they manage to prove that in this situation that an optimal a priori error estimation can be obtained, with a constant that depends only on the shape regularity of the non-degenerate cells. However, these few degenerate simplices deteriorate the condition number of the global stiffness matrix. This can be cured by adding some special terms in the discrete weak form. See [117] for more details.

Last but not least, an unusual technique to generate simplicial meshes is to use residual-based a posteriori estimates, as suggested in [158].

7.8.4   Zoology of A Posteriori Error Estimates

Since the pioneering works of the end of the 1970s, many classes of a posteriori error estimators have been studied. We mention below a few references, for instance, [35, 63, 75, 125, 128], for the interested reader. Some numerical comparisons can be found in e.g. [64, 72].

- Error estimates can be designed by gradient recovery, following the pioneering work [250].
- Error estimates based on equilibrated fluxes allow to have upper bound with explicit constants; see, for instance, [106, 185] for the earliest contributions and [32, 130, 214, 215] for more recent works.
- Functional error estimates allow upper bounds with explicit constants; see [225–227].
- Hierarchical error estimates are based on solving local problems on each simplex using a locally richer discrete space; see, for instance, [28, 64].

For goal-oriented error estimation, there is also an abundant litterature on the topic now. For recent reviews, see especially [124] as well as [35, 75, 116]. For applications to some engineering problems, see, for instance, [62, 154, 229].

More About Practical Implementation

This appendix corresponds to the *Solution Method or Algorithm* of the pipeline below.

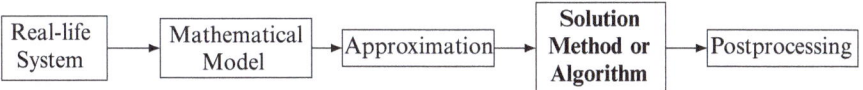

It provides the complete script in scikit-fem [159] to solve Poisson's problem with Nitsche method; see also [79]. The corresponding jupyter notebook is available at:

https://doi.org/10.6084/m9.figshare.25810237.v1

```
1   # Mesh size
2   ref_level = 4
3
4   from skfem import *
5   from skfem.helpers import grad, dot, jump
6   from skfem.models.poisson import laplace, unit_load, mass
7
8   import numpy as np
9
10  # mesh and P1 lagrange finite elements
11  m = MeshTri.init_sqsymmetric().refined(ref_level)
12  e = ElementTriP1()
13  # the Nitsche parameter
14  gamma0N = 1e2
15
16  # define the exact solution
17  def exactsol(x,y):
18      return np.sin(np.pi*x)*np.sin(np.pi*y)
```

F. Chouly, *Finite Element Approximation of Boundary Value Problems*,
Compact Textbooks in Mathematics, https://doi.org/10.1007/978-3-031-72530-2

```
19   # define a source term for solution sin(pi x)sin(pi y)
20   def source(x):
21       return np.pi ** 2 * 2 * ( np.sin(np.pi*x[0])*np.sin(np.pi*x[1]) )
22
23   # interior basis and boundary facet basis
24   ib = Basis(m, e)
25   gb = BoundaryFacetBasis(m,e)
26
27   @LinearForm
28   def rhs(v, _):
29       return source(_.x) * v
30
31   # bilinear form for the penalty term
32   @BilinearForm
33   def penaltyform(u, v, p):
34       h = p.h
35       return gamma0N * u * v / h
36
37   # bilinear form for the Nitsche complementary term
38   @BilinearForm
39   def nitscheform(u, v, p):
40       n = p.n
41       return - dot(grad(u), n) * v - dot(grad(v), n) * u
42
43   # assembly of all the terms
44   A = asm(laplace, ib)
45   N = asm(nitscheform, gb)
46   P = asm(penaltyform, gb)
47   b = rhs.assemble(ib)
48
49   x = solve(A + N + P , b)
50
51   # visualisation
52   M, X = ib.refinterp(x, 4)
53
54   def visualize():
55       from skfem.visuals.matplotlib import plot, draw
56       ax = draw(M, boundaries_only=True)
57       return plot(M, X, shading="gouraud", ax=ax, colorbar=True)
58
59   if __name__ == "__main__":
60       visualize().show()
61
62   # compute the errors
63   M = asm(mass, ib)
```

```
64   u_exact = exactsol(*ib.doflocs[::-1])
65   u_error = x - u_exact
66   error_L2n = np.sqrt(u_error @ M @ u_error)
67   error_H1n = np.sqrt(u_error @ A @ u_error)
68
69   print('L2 error:', error_L2n)
70   print('H1 error (semi-norm):', error_H1n)
71
```

Solutions

Problems of Chap. 2

2.1 We suppose that $k(x) = 1$ for every x in the domain Ω. We use the relationship $\operatorname{div}(\nabla u) = \Delta u$, and we recover Poisson's problem (2.8) with homogeneous Dirichlet boundary condition.

2.2 Let v be a test function on Ω. From (2.34), we get, after the application of the Green formula

$$\int_\Omega k \nabla u \cdot \nabla v - \int_\Gamma (\nabla u \cdot n)\, v = \int_\Omega f v.$$

Then we take v that vanishes on the boundary Γ (Dirichlet boundary condition), which allows to make the boundary term disappear. Then we get the weak form

$$a(u, v) = L(v), \quad \forall v \in V,$$

with

$$a(u, v) = \int_\Omega k \nabla u \cdot \nabla v,$$

$$L(v) = \int_\Omega f v.$$

We take $V = H_0^1(\Omega)$. Thus, for $u, v \in V$ and $f \in L^2(\Omega)$, all the above integrals are meaningful, since the function k is bounded above by k_M.

2.3 First $a(\cdot, \cdot)$ and $L(\cdot)$ are continuous on V. This is a consequence of the bound $k(x) \le k_M$, followed by the Cauchy-Schwarz inequality. Then, from the Poincaré inequality, there holds

$$\|v\|_{1,\Omega}^2 = \|v\|_{0,\Omega}^2 + \|\nabla v\|_{0,\Omega}^2 \le C \|\nabla v\|_{0,\Omega}^2,$$

F. Chouly, *Finite Element Approximation of Boundary Value Problems*,
Compact Textbooks in Mathematics, https://doi.org/10.1007/978-3-031-72530-2

with $C > 0$. We combine the above inequality with the bound $k(x) \geq k_m$ and get:

$$a(v, v) \geq k_m \int_\Omega \nabla v \cdot \nabla v \geq C k_m \|v\|_{1,\Omega}^2,$$

which show that $a(\cdot, \cdot)$ is elliptic (coercive) on V. Existence and uniqueness for the solution to Problem (2.34) follows by application of the Riesz-Fréchet representation Theorem.

2.4 For $v \in H_0^1(\Omega)$ and from (2.35) combined with the Green formula, we get first:

$$\int_{\Omega_1} k_1 \nabla u_1 \cdot \nabla v - \int_\Sigma (\nabla u_1 \cdot n_1) v = \int_{\Omega_1} f v,$$

and then:

$$\int_{\Omega_2} k_2 \nabla u_2 \cdot \nabla v - \int_\Sigma (\nabla u_2 \cdot n_2) v = \int_{\Omega_2} f v.$$

We finally sum the two identities.

2.5 Equation (2.36) combined with the previous result yields:

$$\sum_{i=1}^2 \int_{\Omega_i} k_i \nabla u_i \cdot \nabla v = \sum_{i=1}^2 \int_{\Omega_i} f v.$$

Problems of Chap. 3

3.1 Remark first that any function $v^h \in V^h$ is a continuous, piecewise-polynomial function, so it belongs to $L^2(\Omega)$ and is a regular distribution. Let us compute its distributional gradient, ∇v^h. We take $\varphi \in \mathscr{D}(\Omega; \mathbb{R}^d)$ a test function and compute:

$$\langle \nabla v^h, \varphi \rangle = -\langle v^h, \mathrm{div}\, \varphi \rangle = -\int_\Omega v^h (\mathrm{div}\, \varphi)$$

$$= -\sum_{T \in \mathscr{T}^h} \int_T v^h (\mathrm{div}\, \varphi),$$

where we used the definition of the distributional gradient, the fact that v^h is a regular distribution, and where we split the integral into a sum over the simplices. Since in each simplex T, v^h and φ are smooth functions, we can apply the Green formula:

$$-\int_T v^h (\mathrm{div}\, \varphi) = \int_T \nabla v^h \cdot \varphi - \int_{\partial T} v^h (\varphi \cdot n_T),$$

where n_T is the outward unit normal vector to ∂T. When we sum once again on all the simplices, the boundary terms cancel, because φ is equal to 0 near Γ, and because it is smooth, so that, for each interior edge (or face), the two contributions from adjacent simplices are of opposite sign. Therefore

$$\langle \nabla v^h, \varphi \rangle = \sum_{T \in \mathcal{T}^h} \int_T \nabla v^h \cdot \varphi$$

which means that ∇v^h is a regular distribution: a piecewise polynomial function, discontinuous at the interfaces between simplices, that belongs to $L^2(\Omega)$. We proved that $v^h \in H^1(\Omega)$.

3.2 Take $v \in \mathscr{C}(\overline{\Omega}) \cap H^1(\Omega)$, then we verify:

$$\Upsilon(\mathscr{I}^h v) = \Upsilon \left(\sum_{i=1,\dots,N} v(a_i) \varphi_i \right) = \sum_{i=1,\dots,N} v(a_i) \, \Upsilon \varphi_i$$

$$= \sum_{i \in B} (\Upsilon v)(a_i) \, \Upsilon \varphi_i = \mathscr{I}_\Gamma^h(\Upsilon v),$$

where we used the linearity of the trace operator Υ and the fact that a nodal basis function φ_i vanishes on Γ if its corresponding node a_i is an interior node of the domain Ω.

Problems of Chap. 4

4.1 Suppose that we have the condition: $\exists C \in \mathbb{R}, u = C$ on ∂K. As a result, since

$$m(u) = \frac{C|\partial K|}{|\partial K|} = C,$$

we verify that $u = m(u)$ on ∂K. Conversely, if we suppose that

$$u = m(u) \text{ on } \partial K,$$

then there holds $u = C$ on ∂K, with $C = m(u)$.

4.2 We observe first that V is a vector subspace of $H^1(\Omega \setminus K)$. Let us show that V is a closed subspace. Let us write for this purpose that $V = V_1 \cap V_2$, with

$$V_1 = \{v \in H^1(\Omega \setminus K) \mid v = 0 \text{ on } \Gamma\}, \quad V_2 = \{v \in H^1(\Omega \setminus K) \mid v = m(v) \text{ on } \partial K\}.$$

Let us show that the linear mapping

$$\gamma_K : H^1(\Omega \setminus K) \ni v \mapsto (v - m(v)) \in L^2(\partial K)$$

is bounded. Indeed, with the triangular inequality, and the property that $m(v)$ is constant over ∂K, there holds for $v \in H^1(\Omega \setminus K)$:

$$\|v - m(v)\|_{0,\partial K} \leq \|v\|_{0,\partial K} + \|m(v)\|_{0,\partial K}$$

$$\leq \|v\|_{0,\partial K} + |\partial K|^{\frac{1}{2}}|m(v)|.$$

Moreover we can bound

$$|m(v)| \leq \frac{1}{|\partial K|} \int_{\partial K} |v| \leq \frac{1}{|\partial K|^{\frac{1}{2}}} \|v\|_{0,\partial K}$$

where we used the Cauchy-Schwarz inequality. We use the continuity of the trace operator on ∂K to deduce that

$$\|v - m(v)\|_{0,\partial K} \leq 2\|v\|_{1,\Omega \setminus K}.$$

As a result, V_2 is a closed subspace of $H^1(\Omega \setminus K)$, as V_2 is the kernel of γ_K, a bounded linear mapping. A similar but simpler argument allows to conclude that V_1 is also a closed subspace of $H^1(\Omega \setminus K)$. Then V is a closed subspace of $H^1(\Omega \setminus K)$, as the intersection of two closed subspaces V_1 and V_2. As a result, V is a Hilbert space.

4.3 Let u be a solution to Problem (4.12). Let v be a test function, such that $v = 0$ on Γ. We start from the equation $-\Delta u = f$ on the domain $\Omega \setminus K$, apply the Green formula, take into account the Dirichlet boundary condition $v = 0$ on Γ and get:

$$\int_{\Omega \setminus K} \nabla u \cdot \nabla v - \int_{\partial K} \frac{\partial u}{\partial n} v = \int_{\Omega \setminus K} fv.$$

We write $v = (v - m(v)) + m(v)$ and since $m(v)$ is constant over ∂K, we get then:

$$\int_{\Omega \setminus K} \nabla u \cdot \nabla v - \int_{\partial K} \frac{\partial u}{\partial n}(v - m(v)) - m(v) \int_{\partial K} \frac{\partial u}{\partial n} = \int_{\Omega \setminus K} fv.$$

We take into account the condition $\int_{\partial K} \frac{\partial u}{\partial n} = 0$ for the inclusion, and $v \in V$, and this yields

$$\int_{\Omega \setminus K} \nabla u \cdot \nabla v = \int_{\Omega \setminus K} fv.$$

This is the weak form (4.13).

4.4 Let us take now u a solution to the weak form (4.13). First, since $u \in V$, the second and third equations of Problem (4.12) are verified. Now, let us take $v \in V$ a test function. We use the Green formula and the condition $v = 0$ on Γ, and we transform (4.13) as:

$$-\int_{\Omega \setminus K} (\Delta u)v + \int_{\partial K} \frac{\partial u}{\partial n} v = \int_{\Omega \setminus K} f v.$$

We chose test functions v that vanish on ∂K, and from the Variational Lemma, we get:

$$-\Delta u = f$$

almost everywhere in $\Omega \setminus K$ (the first equation of (4.12) is then satisfied). With the two above equations, we get, for $v \in V$:

$$\int_{\partial K} \frac{\partial u}{\partial n} v = 0.$$

Now with any $v \in V$ such that $m(v) = 1$ on ∂K, we get finally the fourth equation of (4.12):

$$\int_{\partial K} \frac{\partial u}{\partial n} = 0.$$

So we checked that u is also solution to (4.12). The unknown C is in fact $C = m(u)$.

4.5 From the Deny-Lions Lemma, we deduce that

$$v \mapsto \|\nabla v\|_{0,\Omega \setminus K}$$

is a norm over V, equivalent to the norm $\| \cdot \|_{1,\Omega \setminus K}$. As a result, the symmetric bilinear form $a(\cdot, \cdot)$ is coercive on V. Since V is a Hilbert space with $a(\cdot, \cdot)$ bounded (for the H^1-norm) and $L(\cdot)$ bounded (for the H^1-norm), then the Problem (4.13) is well-posed by application of the Riesz-Fréchet representation theorem.

4.6 Since $a(\cdot, \cdot)$ is symmetric, the unique solution u to (4.13) is also the unique minimizer on V of the functional $\mathcal{J} : V \to \mathbb{R}$ which expression is:

$$\mathcal{J}(v) = \frac{1}{2} a(v, v) - L(v).$$

4.7 Let $u_\varepsilon \in H^1(\Omega \setminus K)$ be a minimiser of $\mathcal{J}_\varepsilon$. Let us write the corresponding first order optimality condition:

$$\mathcal{J}'_\varepsilon(u_\varepsilon; v) = 0, \qquad \forall v \in H^1(\Omega \setminus K)$$

with $\mathcal{J}'_\varepsilon$ the differential of $\mathcal{J}_\varepsilon$. We remember that the differential of a quadratic form $\phi(u, u)$ is $2\phi(u, v)$, and that the differential of a linear form is the linear form itself. As a result, when we detail the expression of $\mathcal{J}'_\varepsilon$, we obtain that u_ε is solution to:

$$a(u_\varepsilon, v) + \frac{1}{\varepsilon} \int_\Gamma u_\varepsilon v + \frac{1}{\varepsilon} \int_{\partial K} (u_\varepsilon - m(u_\varepsilon))(v - m(v)) = L(v), \qquad \text{(A.1)}$$

for all $v \in H^1(\Omega \setminus K)$.

4.8 By the Deny-Lions Theorem, the mapping

$$v \mapsto \left(\|\nabla v\|^2_{0,\Omega \setminus K} + \frac{1}{\varepsilon} \int_\Gamma v^2 + \frac{1}{\varepsilon} \int_{\partial K} (v - m(v))^2 \right)^{\frac{1}{2}}$$

defines a norm over $H^1(\Omega \setminus K)$, equivalent to the norm $\| \cdot \|_{1,\Omega \setminus K}$. As a result, the weak form (A.1) admits one unique solution $u_\varepsilon \in H^1(\Omega \setminus K)$, that is also the unique minimizer of $\mathcal{J}_\varepsilon$, by application of the Riesz-Fréchet representation Theorem.

4.9 The discrete formulation that is counterpart of (A.1) consists in finding $u_{\varepsilon,h} \in V_h$ solution to

$$a(u_{\varepsilon,h}, v_h) + \frac{1}{\varepsilon} \int_\Gamma u_{\varepsilon,h} v_h + \frac{1}{\varepsilon} \int_{\partial K} (u_{\varepsilon,h} - m(u_{\varepsilon,h}))(v_h - m(v_h)) = L(v_h), \qquad \text{(A.2)}$$

for all $v_h \in V_h$. This problem admits one unique solution by application of the Riesz-Fréchet representation theorem, since $V_h \subset H^1(\Omega \setminus K)$.

4.10 Let $\varphi_1, \ldots, \varphi_N$ be the global basis of V_h (Galerkin or finite element basis). We write $u_{\varepsilon,h}$ as a linear combination of all the φ_j, and we take as a test function $v_h = \varphi_i$, for $i = 1, \ldots, N$, and we verify that te discrete formulation (A.2) can be rewritten in the following matrix form:

$$\left(\mathbf{K} + \frac{1}{\varepsilon} \mathbf{M}_\Gamma + \frac{1}{\varepsilon} \mathbf{M}_{\partial K} \right) \mathbf{U}_\varepsilon = \mathbf{F}.$$

The coefficients of $\mathbf{K}$ are given by:

$$K_{ij} = \int_{\Omega \setminus K} \nabla \varphi_i \cdot \nabla \varphi_j, \qquad i, j = 1, \ldots, N.$$

The coefficients of $\mathbf{M}_\Gamma$ are

$$M_{\Gamma,ij} = \int_\Gamma \varphi_i \varphi_j, \qquad i, j = 1, \ldots, N.$$

The coefficients of $\mathbf{M}_{\partial K}$ are

$$M_{\partial K,ij} = \int_{\partial K} (\varphi_i - m(\varphi_i))(\varphi_j - m(\varphi_j)), \qquad i, j = 1, \ldots, N.$$

The column vector **F** has its components that read:

$$F_i = \int_{\Omega \setminus K} f \varphi_i, \qquad i = 1, \ldots, N.$$

From the previous results, we deduce that the global matrix $\mathbf{K} + \frac{1}{\varepsilon}\mathbf{M}_\Gamma + \frac{1}{\varepsilon}\mathbf{M}_{\partial K}$ is symmetric positive definite. As a result, it is invertible for any value of $\varepsilon > 0$.

Problems of Chap. 5

5.1 For v_1 and v_2 two test functions in each subdomain Ω_1 and Ω_2 and from (2.35) combined with the Green formula, we get

$$\int_{\Omega_1} k_1 \nabla u_1 \cdot \nabla v_1 - \int_\Sigma \sigma_1 v_1 = \int_{\Omega_1} f v_1,$$

and also

$$\int_{\Omega_2} k_2 \nabla u_2 \cdot \nabla v_2 - \int_\Sigma \sigma_2 v_2 = \int_{\Omega_2} f v_2.$$

We use the interface condition (2.36) on fluxes that yields:

$$\sigma_1 = \frac{1}{2}\sigma_1 + \frac{1}{2}\sigma_1 = \frac{1}{2}\sigma_1 - \frac{1}{2}\sigma_2 = -\frac{1}{2}\sigma_2 - \frac{1}{2}\sigma_2 = -\sigma_2.$$

As a result we define:

$$\langle \sigma(u) \rangle = \frac{1}{2}\sigma_1 - \frac{1}{2}\sigma_2$$

and get from the previous identities

$$\sum_i \int_{\Omega_i} k_i \nabla u_i \cdot \nabla v_i - \int_\Sigma \langle \sigma(u) \rangle [\![v]\!] = \sum_i \int_{\Omega_i} f v_i,$$

For $\theta \in \mathbb{R} \; \gamma > 0$, we rewrite the transmission condition (2.36) on the solution as:

$$\int_\Sigma [\![u]\!](\gamma [\![v]\!] - \theta \langle \sigma(v) \rangle) = 0.$$

A Nitsche method is obtained if we sum the two above identities and use discrete spaces for the trial and test functions. See for instance [38] for the symmetric Nitsche method and [170] for θ arbitrary.

5.2 Well-posedness and optimal convergence are obtained in the same manner as for a Dirichlet condition. See [38] and [142] for details.

Bibliography

1. Abbas, M., Drouet, G., Hild, P.: The local average contact (LAC) method. Comput. Methods Appl. Mech. Eng. **339**, 488–513 (2018). https://doi.org/10.1016/j.cma.2018.05.013
2. Achdou, Y., Deheuvels, T.: A transmission problem across a fractal self-similar interface. Multiscale Model. Simul. **14**(2), 708–736 (2016). https://doi.org/10.1137/15M1029497
3. Achdou, Y., Deheuvels, T., Tchou, N.: $JLip$ versus Sobolev spaces on a class of self-similar fractal foliages. J. Math. Pures Appl. (9) **97**(2), 142–172 (2012). https://doi.org/10.1016/j.matpur.2011.07.002
4. Achdou, Y., Tchou, N.: Trace theorems for a class of ramified domains with self-similar fractal boundaries. SIAM J. Math. Anal. **42**(4), 1449–1482 (2010). https://doi.org/10.1137/090747294
5. Adams, R.A.: Sobolev Spaces. Pure and Applied Mathematics, vol. 65. Academic Press, New York-London (1975)
6. Ainsworth, M., Oden, J.T.: A Posteriori Error Estimation in Finite Element Analysis. Pure and Applied Mathematics: A Wiley Series of Texts, Monographs and Tracts Products Wiley, Chichester (2000)
7. Alart, P., Curnier, A.: A generalized Newton method for contact problems with friction. Journal de Mécanique Théorique et Appliquée **7**(1), 67–82 (1988)
8. Alberty, J., Carstensen, C., Funken, S.A.: Remarks around 50 lines of Matlab: short finite element implementation. Numer. Algorithms **20**(2–3), 117–137 (1999). https://doi.org/10.1023/A:1019155918070
9. Allaire, G.: Numerical Analysis and Optimization. Numerical Mathematics and Scientific Computation. Oxford University Press, Oxford (2007)
10. Allaire, G., Kaber, S.M.: Numerical Linear Algebra. Texts in Applied Mathematics, vol. 55. Springer, New York (2008). https://doi.org/10.1007/978-0-387-68918-0
11. Annavarapu, C., Hautefeuille, M., Dolbow, J.E.: A Nitsche stabilized finite element method for frictional sliding on embedded interfaces. I: single interface. Comput. Methods Appl. Mech. Eng. **268**, 417–436 (2014). https://doi.org/10.1016/j.cma.2013.09.002
12. Apel, T., Milde, F.: Comparison of several mesh refinement strategies near edges. Commun. Numer. Methods Eng. **12**(7), 373–381 (1996). https://doi.org/10.1002/(SICI)1099-0887(199607)12:7<373::AID-CNM985>3.0.CO;2-8
13. Apel, T., Sändig, A.M., Whiteman, J.R.: Graded mesh refinement and error estimates for finite element solutions of elliptic boundary value problems in non-smooth domains. Math. Methods Appl. Sci. **19**(1), 63–85 (1996). https://doi.org/10.1002/(SICI)1099-1476(19960110)19:1<63::AID-MMA764>3.0.CO;2-S
14. Apostolatos, A., Schmidt, R., Wüchner, R., Bletzinger, K.U.: A Nitsche-type formulation and comparison of the most common domain decomposition methods in isogeometric analysis. Internat. J. Numer. Methods Eng. **97**(7), 473–504 (2014). https://doi.org/10.1002/nme.4568

© The Author(s), under exclusive license to Springer Nature Switzerland AG 2025
F. Chouly, *Finite Element Approximation of Boundary Value Problems*,
Compact Textbooks in Mathematics, https://doi.org/10.1007/978-3-031-72530-2

15. Araya, R., Caiazzo, A., Chouly, F.: Stokes problem with slip boundary conditions using stabilized finite elements combined with Nitsche. Comput. Methods Appl. Mech. Eng. **427**, 117037 (2024). https://doi.org/10.1016/j.cma.2024.117037

16. Araya, R., Chouly, F.: Nitsche with a Lagrange Finite Element Method (2023). https://doi.org/10.6084/m9.figshare.24082137.v1. https://figshare.com/articles/software/Nitsche_with_a_Lagrange_Finite_Element_Method/24082137. Figshare repository

17. Araya, R., Chouly, F.: Residual a posteriori error estimation for frictional contact with Nitsche method. J. Sci. Comput. **96**(3) (2023). https://doi.org/10.1007/s10915-023-02300-8

18. Arendt, W., Chalendar, I., Eymard, R.: Galerkin approximation of linear problems in Banach and Hilbert spaces. IMA J. Numer. Anal. **42**(1), 165–198 (2022). https://doi.org/10.1093/imanum/draa067

19. Arnold, D.N.: An interior penalty finite element method with discontinuous elements. SIAM J. Numer. Anal. **19**(4), 742–760 (1982). https://doi.org/10.1137/0719052

20. Arnold, D.N., Falk, R.S., Winther, R.: Finite element exterior calculus, homological techniques, and applications. Acta Numerica **15**, 1–155 (2006). https://doi.org/10.1017/S0962492906210018

21. Astorino, M., Gerbeau, J.F., Pantz, O., Traoré, K.F.: Fluid-structure interaction and multibody contact: application to aortic valves. Comput. Methods Appl. Mech. Eng. **198**(45–46), 3603–3612 (2009). https://doi.org/10.1016/j.cma.2008.09.012

22. Atroshchenko, E., Tomar, S., Xu, G., Bordas, S.P.A.: Weakening the tight coupling between geometry and simulation in isogeometric analysis: from sub- and super-geometric analysis to geometry-independent field approximation (GIFT). Int. J. Numer. Methods Eng. **114**(10), 1131–1159 (2018). https://doi.org/10.1002/nme.5778

23. Baaijens, F.P.T.: A fictitious domain/mortar element method for fluid-structure interaction. Int. J. Numer. Methods Fluids **35**(7), 743–761 (2001). https://doi.org/10.1002/fld.153

24. Babuška, I.: Finite element method for domains with corners. Computing **6**, 264–273 (1970). https://doi.org/10.1007/BF02238811

25. Babuška, I.: The finite element method with Lagrangian multipliers. Numer. Math. **20**, 179–192 (1972/73)

26. Babuška, I.: The finite element method with penalty. Math. Comput. **27**, 221–228 (1973)

27. Bacuta, C., Bramble, J.H., Xu, J.: Regularity estimates for elliptic boundary value problems in Besov spaces. Math. Comput. **72**(244), 1577–1595 (2003). https://doi.org/10.1090/S0025-5718-02-01502-8

28. Bank, R.E., Weiser, A.: Some a posteriori error estimators for elliptic partial differential equations. Math. Comput. **44**, 283–301 (1985). https://doi.org/10.2307/2007953

29. Bansal, A., Barnafi, N.A., Pandey, D.N.: Nitsche method for Navier-Stokes equations with slip boundary conditions: convergence analysis and VMS-LES stabilization (2023). ArXiv eprint 2307.03589. https://doi.org/10.1051/m2an/2024070

30. Barbosa, H.J.C., Hughes, T.J.R.: The finite element method with Lagrange multipliers on the boundary: circumventing the Babuška-Brezzi condition. Comput. Methods Appl. Mech. Eng. **85**(1), 109–128 (1991). https://doi.org/10.1016/0045-7825(91)90125-P

31. Bartels, S., Carstensen, C., Dolzmann, G.: Inhomogeneous Dirichlet conditions in a priori and a posteriori finite element error analysis. Numer. Math. **99**(1), 1–24 (2004). https://doi.org/10.1007/s00211-004-0548-3

32. Bartels, S., Kaltenbach, A.: Exact error control for variational problems via convex duality and explicit flux reconstruction. In: Chouly, F., Bordas, S.P.A., Becker, R., Omnes, P. (eds.) Error Control, Adaptive Discretizations, and Applications, Part 1. Advances in Applied Mechanics (AAMS). ISBN: 9780443294488, vol. 58. Elsevier, Amsterdam (2024). https://doi.org/10.1016/bs.aams.2024.04.001

33. Bazilevs, Y., Hughes, T.J.R.: Weak imposition of Dirichlet boundary conditions in fluid mechanics. Comput. Fluids **36**(1), 12–26 (2007). https://doi.org/10.1016/j.compfluid.2005.07.012.

34. Beaude, L., Chouly, F., Laaziri, M., Masson, R.: Mixed and Nitsche's discretizations of Coulomb frictional contact-mechanics for mixed dimensional poromechanical models.

Comput. Methods Appl. Mech. Eng. **413**, 116124 (2023). https://doi.org/10.1016/j.cma.2023.116124

35. Becker, R., Bordas, S.P.A., Chouly, F., Omnes, P.: A short perspective on a posteriori error control and adaptive discretizations. In: Chouly, F., Bordas, S.P.A., Becker, R., Omnes, P. (eds.) Error Control, Adaptive Discretizations, and Applications, Part 1. Advances in Applied Mechanics (AAMS). ISBN: 9780443294488, vol. 58. Elsevier, Amsterdam (2024). https://doi.org/10.1016/bs.aams.2024.03.002

36. Becker, R., Burman, E., Hansbo, P.: A Nitsche extended finite element method for incompressible elasticity with discontinuous modulus of elasticity. Comput. Methods Appl. Mech. Eng. **198**(41–44), 3352–3360 (2009). https://doi.org/10.1016/j.cma.2009.06.017

37. Becker, R., Estecahandy, E., Trujillo, D.: Weighted marking for goal-oriented adaptive finite element methods. SIAM J Numer. Anal. **49**(6), 2451–2469 (2011)

38. Becker, R., Hansbo, P., Stenberg, R.: A finite element method for domain decomposition with non-matching grids. M2AN Math. Model. Numer. Anal. **37**(2), 209–225 (2003). https://doi.org/10.1051/m2an:2003023

39. Becker, R., Rannacher, R.: An optimal control approach to a posteriori error estimation in finite element methods. Acta Numer. **10**, 1–102 (2001). https://doi.org/10.1017/S0962492901000010

40. Beirão da Veiga, L., Brezzi, F., Cangiani, A., Manzini, G., Marini, L.D., Russo, A.: Basic principles of virtual element methods. Math. Models Methods Appl. Sci. **23**(1), 199–214 (2013). https://doi.org/10.1142/S0218202512500492

41. Ben Belgacem, F.: The mortar finite element method with Lagrange multipliers. Numer. Math. **84**(2), 173–197 (1999). https://doi.org/10.1007/s002110050468

42. Ben Belgacem, F.: Why is the Cauchy problem severely ill-posed? Inverse Probl. **23**(2), 823–836 (2007). https://doi.org/10.1088/0266-5611/23/2/020

43. Ben Belgacem, F., Renard, Y.: Hybrid finite element methods for the Signorini problem. Math. Comput. **72**(243), 1117–1145 (2003). https://doi.org/10.1090/S0025-5718-03-01490-X

44. Bernardi, C., Maday, Y.: Spectral methods. In: Handbook of Numerical Analysis, vol. V, pp. 209–485. North-Holland, Amsterdam (1997)

45. Bernardi, C., Maday, Y., Patera, A.T.: Domain decomposition by the mortar element method. In: Asymptotic and Numerical Methods for Partial Differential Equations with Critical Parameters (Beaune, 1992), NATO Adv. Sci. Inst. Ser. C Math. Phys. Sci., vol. 384, pp. 269–286. Kluwer Academic Publishers, Dordrecht (1993)

46. Bernardi, C., Maday, Y., Patera, A.T.: A new nonconforming approach to domain decomposition: the mortar element method. In: Nonlinear Partial Differential Equations and Their Applications. Collège de France Seminar, vol. XI (Paris, 1989–1991). Pitman Research Notes in Mathematics Series, vol. 299, pp. 13–51. Longman Scientific & Technical Publisher, Harlow (1994)

47. Bersetche, F.M., Borthagaray, J.P.: A deep first-order system least squares method for solving elliptic PDEs. Comput. Math. Appl. **129**, 136–150 (2023). https://doi.org/10.1016/j.camwa.2022.11.014

48. Bertoluzza, S.: A posteriori error estimates for the wavelet Galerkin method. Appl. Math. Lett. **8**(5), 1–6 (1995). https://doi.org/10.1016/0893-9659(95)00057-W

49. Bertoluzza, S., Burman, E.: An abstract framework for heterogeneous coupling: stability, approximation and applications (2023). ArXiv eprint 2312.11733

50. Bertoluzza, S., Naldi, G., Ravel, J.C.: Wavelet methods for the numerical solution of boundary value problems on the interval. In: Wavelets: Theory, Algorithms, and Applications. Proceedings of the International Conference on Wavelets, Held in Taormina, Italy, October 14–20, 1993, pp. 425–448. Academic Press, San Diego (1994)

51. Bochev, P., Lehoucq, R.B.: On the finite element solution of the pure Neumann problem. SIAM Rev. **47**(1), 50–66 (2005). https://doi.org/10.1137/S0036144503426074

52. Boffi, D., Cangiani, A., Feder, M., Gastaldi, L., Heltai, L.: A comparison of non-matching techniques for the finite element approximation of interface problems. Comput. Math. Appl. **151**, 101–115 (2023). https://doi.org/10.1016/j.camwa.2023.09.017

53. Bonnaillie-Noël, V., Dambrine, M., Hérau, F., Vial, G.: On generalized Ventcel's type boundary conditions for Laplace operator in a bounded domain. SIAM J. Math. Anal. **42**(2), 931–945 (2010). https://doi.org/10.1137/090756521

54. Bordas, S.P.A., Menk, A.: Partition of Unity Methods. Wiley, London (2023)

55. Bourlard, M., Dauge, M., Lubuma, M.S., Nicaise, S.: Coefficients of the singularities for elliptic boundary value problems on domains with conical points. III. Finite element methods on polygonal domains. SIAM J. Numer. Anal. **29**(1), 136–155 (1992). https://doi.org/10.1137/0729009

56. Braack, M., Ern, A.: A posteriori control of modeling errors and discretization errors. Multiscale Model. Simul. **1**(2), 221–238 (2003). https://doi.org/10.1137/S1540345902410482

57. Brandts, J., Korotov, S., Křížek, M.: On the equivalence of regularity criteria for triangular and tetrahedral finite element partitions. Comput. Math. Appl. **55**(10), 2227–2233 (2008). https://doi.org/10.1016/j.camwa.2007.11.010

58. Brenner, S.C.: Multigrid methods for the computation of singular solutions and stress intensity factors. I: Corner singularities. Math. Comput. **68**(226), 559–583 (1999). https://doi.org/10.1090/S0025-5718-99-01017-0

59. Brenner, S.C., Scott, L.R.: The Mathematical Theory of Finite Element Methods. Texts in Applied Mathematics, vol. 15, 3rd edn. Springer, New York (2008). https://doi.org/10.1007/978-0-387-75934-0

60. Brezis, H.: Analyse fonctionnelle. Collection Mathématiques Appliquées pour la Maîtrise. [Collection of Applied Mathematics for the Master's Degree]. Masson, Paris (1983)

61. Brezzi, F., Fortin, M.: A minimal stabilisation procedure for mixed finite element methods. Numer. Math. **89**(3), 457–491 (2001). https://doi.org/10.1007/PL00005475.

62. Bui, H.P., Duprez, M., Rohan, P.Y., Lejeune, A., Bordas, S.P.A., Bucki, M., Chouly, F.: Enhancing biomechanical simulations based on a posteriori error estimates: the potential of Dual Weighted Residual-driven adaptive mesh refinement (2024). ArXiv eprint 2403.00401

63. Bulle, R.: A posteriori error estimation for finite element approximations of fractional Laplacian problems and applications to poro-elasticity. Theses, Université Bourgogne Franche-Comté ; Université du Luxembourg (2022). https://theses.hal.science/tel-03652547

64. Bulle, R., Hale, J.S., Lozinski, A., Bordas, S.P.A., Chouly, F.: Hierarchical a posteriori error estimation of Bank-Weiser type in the FEniCS Project. Comput. Math. Appl. **131**, 103–123 (2023). https://doi.org/10.1016/j.camwa.2022.11.009

65. Burman, E.: A penalty-free nonsymmetric Nitsche-type method for the weak imposition of boundary conditions. SIAM J. Numer. Anal. **50**(4), 1959–1981 (2012). https://doi.org/10.1137/10081784X

66. Burman, E., Claus, S., Hansbo, P., Larson, M.G., Massing, A.: CutFEM: discretizing geometry and partial differential equations. Internat. J. Numer. Methods Eng. **104**(7), 472–501 (2015). https://doi.org/10.1002/nme.4823

67. Burman, E., Fernández, M.A., Frei, S.: A Nitsche-based formulation for fluid-structure interactions with contact. ESAIM Math. Model. Numer. Anal. **54**(2), 531–564 (2020). https://doi.org/10.1051/m2an/2019072

68. Burman, E., Hansbo, P.: Fictitious domain finite element methods using cut elements: II. A stabilized Nitsche method. Appl. Numer. Math. **62**(4), 328–341 (2012). https://doi.org/10.1016/j.apnum.2011.01.008

69. Burman, E., Hansbo, P., Larson, M.G.: The augmented Lagrangian method as a framework for stabilised methods in computational mechanics. Arch. Comput. Methods Eng. **30**(4), 2579–2604 (2023)

70. Burman, E., Hansbo, P., Larson, M.G., Stenberg, R.: Galerkin least squares finite element method for the obstacle problem. Comput. Methods Appl. Mech. Eng. **313**, 362–374 (2017). https://doi.org/10.1016/j.cma.2016.09.025

71. Canuto, C., Hussaini, M.Y., Quarteroni, A., Zang, T.A.: Spectral Methods. Scientific Computation. Springer, Berlin (2006)

72. Carstensen, C., Merdon, C.: Estimator competition for Poisson problems. J. Comput. Math. **28**(3), 309–330 (2010). https://doi.org/10.4208/jcm.2009.10-m1015

73. Cascavita, K.L., Chouly, F., Ern, A.: Hybrid high-order discretizations combined with Nitsche's method for Dirichlet and Signorini boundary conditions. IMA J. Numer. Anal. **40**(4), 2189–2226 (2020). https://doi.org/10.1093/imanum/drz038

74. Chahine, E., Laborde, P., Renard, Y.: Crack tip enrichment in the XFEM using a cutoff function. Int. J. Numer. Methods Eng. **75**(6), 629–646 (2008). https://doi.org/10.1002/nme.2265

75. Chamoin, L., Legoll, F.: An introductory review on a posteriori error estimation in finite element computations. SIAM Rev. **65**(4), 963–1028 (2023)

76. Chapelle, D., Bathe, K.J.: The finite element analysis of shells—fundamentals, 2nd edn. Computational Fluid and Solid Mechanics. Springer, Heidelberg (2011). https://doi.org/10.1007/978-3-642-16408-8

77. Chen, T., Mo, R., Wan, N., Gong, Z.W.: Imposing displacement boundary conditions with Nitsche's method in isogeometric analysis. Chin. J. Theor. Appl. Mech. **44**(2), 369–381 (2012)

78. Chouly, F.: A review on some discrete variational techniques for the approximation of essential boundary conditions. Viet. J. Math. (2024). https://link.springer.com/article/10.1007/s10013-024-00702-1

79. Chouly, F.: scikit-fem jupyter notebook for Nitsche method (2024). https://doi.org/10.6084/m9.figshare.25810237.v1. https://figshare.com/articles/dataset/scikit-fem_jupyter_notebook_for_Nitsche_method/25810237

80. Chouly, F., Ern, A., Pignet, N.: A hybrid high-order discretization combined with Nitsche's method for contact and Tresca friction in small strain elasticity. SIAM J. Sci. Comput. **42**(4), A2300–A2324 (2020). https://doi.org/10.1137/19M1286499

81. Chouly, F., Fabre, M., Hild, P., Mlika, R., Pousin, J., Renard, Y.: An overview of recent results on Nitsche's method for contact problems. In: Geometrically Unfitted Finite Element Methods and Applications, Lecture Notes in Computational Science and Engineering, vol. 121, pp. 93–141. Springer, Cham (2017). https://doi.org/10.1007/978-3-319-71431-8_4

82. Chouly, F., Fabre, M., Hild, P., Pousin, J., Renard, Y.: Residual-based a posteriori error estimation for contact problems approximated by Nitsche's method. IMA J. Numer. Anal. **38**(2), 921–954 (2018). https://doi.org/10.1093/imanum/drx024

83. Chouly, F., Gustafsson, T., Hild, P.: A Nitsche method for the elastoplastic torsion problem. ESAIM Math. Model. Numer. Anal. **57**(3), 1731–1746 (2023). https://doi.org/10.1051/m2an/2023034

84. Chouly, F., Hild, P.: A Nitsche-based method for unilateral contact problems: numerical analysis. SIAM J. Numer. Anal. **51**(2), 1295–1307 (2013). https://doi.org/10.1137/12088344X

85. Chouly, F., Hild, P., Lleras, V., Renard, Y.: Nitsche method for contact with Coulomb friction: existence results for the static and dynamic finite element formulations. J. Comput. Appl. Math. **416**, 18 (2022). https://doi.org/10.1016/j.cam.2022.114557

86. Chouly, F., Hild, P., Renard, Y.: Finite element approximation of contact and friction in elasticity. In: Advances in Mechanics and Mathematics/Advances in Continuum Mechanics, vol. 48. Springer, Birkhäuser (2023). ISBN 978-3-031-31422-3. https://doi.org/10.1007/978-3-031-31423-0, xxi+294 p.

87. Ciarlet, P.G.: The finite element method for elliptic problems, In: Classics in Applied Mathematics, vol. 40. Society for Industrial and Applied Mathematics (SIAM), Philadelphia (2002). https://doi.org/10.1137/1.9780898719208

88. Ciarlet, P.G., Lions, J.L. (eds.) Finite Element Methods (Part 1). Handbook of Numerical Analysis, vol. 2. North-Holland, Amsterdam (1991)

89. Cicuttin, M., Ern, A., Pignet, N.: Hybrid high-order methods—a primer with applications to solid mechanics. SpringerBriefs in Mathematics. Springer, Cham (2021). https://doi.org/10.1007/978-3-030-81477-9

90. Cockburn, B., Di Pietro, D.A., Ern, A.: Bridging the hybrid high-order and hybridizable discontinuous Galerkin methods. ESAIM Math. Model. Numer. Anal. **50**(3), 635–650 (2016). https://doi.org/10.1051/m2an/2015051

91. Cohen, A., Dahmen, W., DeVore, R.: Adaptive wavelet methods for elliptic operator equations: convergence rates. Math. Comput. **70**(233), 27–75 (2001). https://doi.org/10.1090/S0025-5718-00-01252-7

92. Cohen, A., Masson, R.: Wavelet adaptive method for second order elliptic problems: Boundary conditions and domain decomposition. Numer. Math. **86**(2), 193–238 (2000). https://doi.org/10.1007/s002110000158

93. Collatz, L.: Funktionalanalysis und numerische Mathematik, Grundlehren der mathematischen Wissenschaften, vol. 120. Springer, Berlin-Göttingen-Heidelberg (1964)

94. Costabel, M., Dauge, M.: Crack singularities for general elliptic systems. Math. Nachr. **235**, 29–49 (2002). https://doi.org/10.1002/1522-2616(200202)235:1<29::AID-MANA29>3.0.CO;2-6

95. Cottrell, J.A., Hughes, T.J.R., Bazilevs, Y.: Isogeometric Analysis. John Wiley & Sons, Chichester (2009). https://doi.org/10.1002/9780470749081.

96. Dahlke, S.: Besov regularity for elliptic boundary value problems in polygonal domains. Appl. Math. Lett. **12**(6), 31–36 (1999). https://doi.org/10.1016/S0893-9659(99)00075-0

97. Dahlke, S.: Besov regularity of edge singularities for the Poisson equation in polyhedral domains. Numer. Linear Algebra Appl. **9**(6–7), 457–466 (2002). https://doi.org/10.1002/nla.304

98. Dahlke, S., DeVore, R.A.: Besov regularity for elliptic boundary value problems. Commun. Partial Differ. Equ. **22**(1–2), 1–16 (1997). https://doi.org/10.1080/03605309708821252

99. Dauge, M.: Elliptic Boundary Value Problems on Corner Domains. Smoothness and Asymptotics of Solutions. Lecture Notes in Mathematics, vol. 1341. Springer, Berlin (1988). https://doi.org/10.1007/BFb0086682

100. Dauge, M., Nicaise, S., Bourlard, M., Lubuma, J.M.S.: Coefficients des singularités pour des problèmes aux limites elliptiques sur un domaine à points coniques. I. Résultats généraux pour le problème de Dirichlet. RAIRO Modél. Math. Anal. Numér. **24**(1), 27–52 (1990). https://doi.org/10.1051/m2an/1990240100271

101. Dauge, M., Nicaise, S., Bourlard, M., Lubuma, J.M.S.: Coefficients des singularités pour des problèmes aux limites elliptiques sur un domaine à points coniques. II. Quelques opérateurs particuliers. RAIRO Modél. Math. Anal. Numér. **24**(3), 343–367 (1990). https://doi.org/10.1051/m2an/1990240303431

102. Dautray, R., Lions, J.L.: Mathematical analysis and numerical methods for science and technology. Volume 1: Physical origins and classical methods. With the collaboration of Philippe Bénilan, Michel Cessenat, André Gervat, Alain Kavenoky, Hélène Lanchon. Transl. from the French by Ian N. Sneddon., 2nd printing edn. Springer, Berlin (2000)

103. Dekel, S., Leviatan, D.: Whitney estimates for convex domains with applications to multivariate piecewise polynomial approximation. Found. Comput. Math. **4**(4), 345–368 (2004). https://doi.org/10.1007/s10208-004-0096-3

104. Demkowicz, L., Gopalakrishnan, J.: A class of discontinuous Petrov-Galerkin methods. Part I: the transport equation. Comput. Methods Appl. Mech. Eng. **199**(23–24), 1558–1572 (2010). https://doi.org/10.1016/j.cma.2010.01.003

105. Demkowicz, L., Gopalakrishnan, J., Niemi, A.H.: A class of discontinuous Petrov-Galerkin methods. Part III: Adaptivity. Appl. Numer. Math. **62**(4), 396–427 (2012). https://doi.org/10.1016/j.apnum.2011.09.002

106. Destuynder, P., Djaoua, M.: Estimation de l'erreur sur le coefficient de la singularite de la solution d'un problème elliptique sur un ouvert avec coin. RAIRO, Anal. Numér. **14**, 239–248 (1980)

107. Di Nezza, E., Palatucci, G., Valdinoci, E.: Hitchhiker's guide to the fractional Sobolev spaces. Bull. Sci. Math. **136**(5), 521–573 (2012). https://doi.org/10.1016/j.bulsci.2011.12.004

108. Di Pietro, D.A., Droniou, J.: The hybrid high-order method for polytopal meshes, *MS&A. Modeling, Simulation and Applications*, vol. 19. Springer, Cham (2020). https://doi.org/10.1007/978-3-030-37203-3

109. Di Pietro, D.A., Ern, A.: Mathematical Aspects of Discontinuous Galerkin Methods. Mathematics & Applications, vol. 69. Springer, Heidelberg (2012). https://doi.org/10.1007/978-3-642-22980-0

110. Di Pietro, D.A., Fontana, I., Kazymyrenko, K.: A posteriori error estimates via equilibrated stress reconstructions for contact problems approximated by Nitsche's method. Comput. Math. Appl. **111**, 61–80 (2022). https://doi.org/10.1016/j.camwa.2022.02.008

111. Dolbow, J., Moës, N., Belytschko, T.: An extended finite element method for modeling crack growth with frictional contact. Comput. Methods Appl. Mech. Eng. **190**(51–52), 6825–6846 (2001). https://doi.org/10.1016/S0045-7825(01)00260-2

112. Dong, Z., Ern, A.: Hybrid high-order and weak Galerkin methods for the biharmonic problem. SIAM J. Numer. Anal. **60**(5), 2626–2656 (2022)

113. Drouet, G., Hild, P.: Optimal convergence for discrete variational inequalities modelling Signorini contact in 2D and 3D without additional assumptions on the unknown contact set. SIAM J. Numer. Anal. **53**(3), 1488–1507 (2015). https://doi.org/10.1137/140980697

114. Duboeuf, F., Béchet, E.: Embedded solids of any dimension in the X-FEM. Part II – Imposing Dirichlet boundary conditions. Finite Elem. Anal. Des. **128**, 32–50 (2017). https://doi.org/https://doi.org/10.1016/j.finel.2017.01.005. https://www.sciencedirect.com/science/article/pii/S0168874X16301974

115. Dupont, T., Scott, R.: Polynomial approximation of functions in Sobolev spaces. Math. Comput. **34**(150), 441–463 (1980). https://doi.org/10.2307/2006095

116. Duprez, M., Bordas, S.P.A., Bucki, M., Bui, H.P., Chouly, F., Lleras, V., Lobos, C., Lozinski, A., Rohan, P.Y., Tomar, S.: Quantifying discretization errors for soft tissue simulation in computer assisted surgery: a preliminary study. Appl. Math. Model. **77**, 709–723 (2020). https://doi.org/10.1016/j.apm.2019.07.055

117. Duprez, M., Lleras, V., Lozinski, A.: Finite element method with local damage of the mesh. ESAIM Math. Model. Numer. Anal. **53**(6), 1871–1891 (2019). https://doi.org/10.1051/m2an/2019023

118. Duprez, M., Lozinski, A.: ϕ-FEM: a finite element method on domains defined by level-sets. SIAM J. Numer. Anal. **58**(2), 1008–1028 (2020)

119. Durán, R.G., López García, F.: Solutions of the divergence and Korn inequalities on domains with an external cusp. Ann. Acad. Sci. Fenn., Math. **35**(2), 421–438 (2010). https://doi.org/10.5186/aasfm.2010.3527

120. Duvaut, G., Lions, J.L.: Les inéquations en mécanique et en physique. Dunod, Paris (1972). Travaux et Recherches Mathématiques, No. 21

121. Eck, C., Jarušek, J., Krbec, M.: Unilateral Contact Problems. Pure and Applied Mathematics (Boca Raton), vol. 270. Chapman & Hall/CRC, Boca Raton (2005). https://doi.org/10.1201/9781420027365

122. Egger, H., Rüde, U., Wohlmuth, B.: Energy-corrected finite element methods for corner singularities. SIAM J. Numer. Anal. **52**(1), 171–193 (2014). https://doi.org/10.1137/120871377

123. Embar, A., Dolbow, J., Harari, I.: Imposing Dirichlet boundary conditions with Nitsche's method and spline-based finite elements. Int. J. Numer. Methods Eng. **83**(7), 877–898 (2010). https://doi.org/10.1002/nme.2863

124. Endtmayer, B., Langer, U., Richter, T., Schafelner, A., Wick, T.: A posteriori single- and multi-goal error control and adaptivity for partial differential equations. In: Chouly, F., Bordas, S.P.A., Becker, R., Omnes, P. (eds.) Error Control, Adaptive Discretizations, and Applications, Part 2. Advances in Applied Mechanics (AAMS), vol. 59. Elsevier, Amsterdam (2024). https://doi.org/10.1016/bs.aams.2024.08.003

125. Ern, A., Guermond, J.L.: Theory and Practice of Finite Elements. Applied Mathematical Sciences, vol. 159. Springer, New York (2004)

126. Ern, A., Guermond, J.L.: Abstract nonconforming error estimates and application to boundary penalty methods for diffusion equations and time-harmonic Maxwell's equations. Comput. Methods Appl. Math. **18**(3), 451–475 (2018). https://doi.org/10.1515/cmam-2017-0058

127. Ern, A., Guermond, J.L.: Finite Elements. I—Approximation and Interpolation. Texts in Applied Mathematics, vol. 72. Springer, Cham (2021). https://doi.org/10.1007/978-3-030-56341-7

128. Ern, A., Guermond, J.L.: Finite Elements II—Galerkin Approximation, Elliptic and Mixed PDEs. Texts in Applied Mathematics, vol. 73. Springer, Cham (2021). https://doi.org/10.1007/978-3-030-56923-5

129. Ern, A., Guermond, J.L.: Finite Elements III—First-Order and Time-Dependent PDEs. Texts in Applied Mathematics, vol. 74. Springer, Cham (2021). https://doi.org/10.1007/978-3-030-57348-5

130. Ern, A., Vohralík, M.: Adaptive inexact Newton methods with a posteriori stopping criteria for nonlinear diffusion PDEs. SIAM J. Sci. Comput. **35**(4), a1761–a1791 (2013). https://doi.org/10.1137/120896918

131. Falk, R.: Error estimates for the approximation of a class of variational inequalities. Math. Comput. **28**(128), 963–971 (1974)

132. Farhat, C., Roux, F.X.: A method of finite element tearing and interconnecting and its parallel solution algorithm. Int. J. Numer. Methods Eng. **32**(6), 1205–1227 (1991). https://doi.org/10.1002/nme.1620320604

133. Federer, H.: Geometric measure theory. Die Grundlehren der mathematischen Wissenschaften, Band 153. Springer, New York (1969)

134. Fernández, M.A.: Coupling schemes for incompressible fluid-structure interaction: implicit, semi-implicit and explicit. SeMA J. **55**(55), 59–108 (2011)

135. Fernández, M.A., Gerbeau, J.F.: Algorithms for fluid-structure interaction problems. In: Cardiovascular Mathematics. MS&A Modeling and Simulation: An Application, vol. 1, pp. 307–346. Springer Italia, Milan (2009). https://doi.org/10.1007/978-88-470-1152-6_9

136. Fontana, I., Di Pietro, D.A.: An a posteriori error analysis based on equilibrated stresses for finite element approximations of frictional contact. Comput. Methods Appl. Mech. Eng. **425**, 26 (2024). https://doi.org/10.1016/j.cma.2024.116950

137. Formaggia, L., Gatti, F., Zonca, S.: An XFEM/DG approach for fluid-structure interaction problems with contact. Appl. Math. **66**(2), 183–211 (2021). https://doi.org/10.21136/AM.2021.0310-19

138. Formaggia, L., Perotto, S.: Anisotropic error estimates for elliptic problems. Numer. Math. **94**(1), 67–92 (2003). https://doi.org/10.1007/s00211-002-0415-z

139. Fortin, A.: An anisotropic mesh adaptation method based on gradient recovery and optimal shape elements. In: Chouly, F., Bordas, S.P.A., Becker, R., Omnes, P. (eds.) Error Control, Adaptive Discretizations, and Applications, Part 1. Advances in Applied Mechanics (AAMS), vol. 58. Elsevier, Amsterdam (2024). https://doi.org/10.1016/bs.aams.2024.03.003

140. Freund, J., Stenberg, R.: On weakly imposed boundary conditions for second order problems. In: Proceedings of the Ninth Int. Conf. Finite Elements in Fluids, pp. 327–336. Venice (1995)

141. Frey, P.J., George, P.L.: Mesh generation. Application to finite elements., 2nd edn. ISTE, London; John Wiley, Hoboken (2008). https://doi.org/10.1002/9780470611166

142. Fritz, A., Hüeber, S., Wohlmuth, B.I.: A comparison of mortar and Nitsche techniques for linear elasticity. Calcolo **41**(3), 115–137 (2004). https://doi.org/10.1007/s10092-004-0087-4

143. Gander, M.J.: Optimized Schwarz methods. SIAM J. Numer. Anal. **44**(2), 699–731 (2006). https://doi.org/10.1137/S0036142903425409

144. Gander, M.J.: Schwarz methods over the course of time. ETNA, Electron. Trans. Numer. Anal. **31**, 228–255 (2008)

145. Gander, M.J.: 50 years of time parallel time integration. In: Multiple shooting and time domain decomposition methods. MuS-TDD, Heidelberg, Germany, May 6–8, 2013, pp. 69–113. Springer, Cham (2015). https://doi.org/10.1007/978-3-319-23321-5_3

146. Gander, M.J., Kwok, F.: Numerical Analysis of Partial Differential Equations Using Maple and MATLAB. Fundam. Algorithms, vol. 12. Society for Industrial and Applied Mathematics (SIAM), Philadelphia (2018). https://doi.org/10.1137/1.9781611975314

147. Gander, M.J., Wanner, G.: From Euler, Ritz, and Galerkin to modern computing. SIAM Rev. **54**(4), 627–666 (2012). https://doi.org/10.1137/100804036

148. Gasquet, C., Witomski, P.: Fourier Analysis and Applications. Texts in Applied Mathematics, vol. 30. Springer, New York (1999). https://doi.org/10.1007/978-1-4612-1598-1

149. Gerstenberger, A., Wall, W.A.: An extended finite element method/Lagrange multiplier based approach for fluid-structure interaction. Comput. Methods Appl. Mech. Eng. **197**(19–20), 1699–1714 (2008). https://doi.org/10.1016/j.cma.2007.07.002

150. Girault, V., Raviart, P.A.: Finite Element Methods for Navier-Stokes Equations. Theory and Algorithms. Springer Series in Computational Mathematics, vol. 5 (Extended version of the 1979 publ.) edn. Springer, Cham (1986). https://doi.org/10.1007/978-3-642-61623-5

151. Gjerde, I.G., Scott, L.R.: Nitsche's method for Navier-Stokes equations with slip boundary conditions. Math. Comput. **91**(334), 597–622 (2022). https://doi.org/10.1090/mcom/3682

152. Glowinski, R.: Numerical Methods for Nonlinear Variational Problems. Springer Series in Computational Physics. Springer, New York (1984). https://doi.org/10.1007/978-3-662-12613-4

153. Glowinski, R., Pan, T., Périaux, J.: A fictitious domain method for Dirichlet problem and applications. Comput. Methods Appl. Mech. Eng. **111**(3–4), 283–303 (1994)

154. González-Estrada, O.A., Nadal, E., Ródenas, J.J., Kerfriden, P., Bordas, S.P.A., Fuenmayor, F.J.: Mesh adaptivity driven by goal-oriented locally equilibrated superconvergent patch recovery. Comput. Mech. **53**(5), 957–976 (2014)

155. Grepl, M.A., Maday, Y., Nguyen, N.C., Patera, A.T.: Efficient reduced-basis treatment of nonaffine and nonlinear partial differential equations. M2AN Math. Model. Numer. Anal. **41**(3), 575–605 (2007)

156. Grisvard, P.: Elliptic Problems in Nonsmooth Domains. Monographs and Studies in Mathematics, vol. 24. Pitman (Advanced Publishing Program), Boston (1985)

157. Grisvard, P.: Problèmes aux limites dans les polygones. Mode d'emploi. EDF Bull. Direction Études Rech. Sér. C Math. Inform. (1), 3, 21–59 (1986)

158. Gustafsson, T.: A simple technique for unstructured mesh generation via adaptive finite elements. Rakenteiden Mekaniikka **54**(2), 69–79 (2021). https://doi.org/10.23998/rm.99648. https://rakenteidenmekaniikka.journal.fi/article/view/99648

159. Gustafsson, T., Mcbain, G.D.: scikit-fem: A Python package for finite element assembly. J. Open Source Softw. **5**(52), 2369 (2020)

160. Gustafsson, T., Stenberg, R., Videman, J.: Mixed and stabilized finite element methods for the obstacle problem. SIAM J. Numer. Anal. **55**(6), 2718–2744 (2017). https://doi.org/10.1137/16M1065422

161. Gustafsson, T., Stenberg, R., Videman, J.: Error analysis of Nitsche's mortar method. Numer. Math. **142**(4), 973–994 (2019). https://doi.org/10.1007/s00211-019-01039-5

162. Gustafsson, T., Stenberg, R., Videman, J.: Nitsche's method for Kirchhoff plates. SIAM J. Sci. Comput. **43**(3), a1651–a1670 (2021). https://doi.org/10.1137/20M1349801

163. Hansbo, P.: Nitsche's method for interface problems in computational mechanics. GAMM-Mitteilungen **28**(2), 183–206 (2005)

164. Haslinger, J., Hlaváček, I., Nečas, J.: Handbook of numerical analysis. In: Ciarlet, P.G., Lions, J.L. (eds.) Numerical Methods for Unilateral Problems in Solid Mechanics, vol. IV, chap. 2, pp. 313–385. North Holland, Amsterdam (1996)

165. Haslinger, J., Renard, Y.: A new fictitious domain approach inspired by the extended finite element method. SIAM J. Numer. Anal. **47**(2), 1474–1499 (2009). https://doi.org/10.1137/070704435

166. Hauret, P.: At the crossroads of simulation and data analytics. Eur. Math. Soc. Mag. **121**, 9–18 (2021). https://doi.org/10.4171/MAG/20

167. Hecht, F.: New development in freefem++. J. Numer. Math. **20**(3–4), 251–265 (2012). https://doi.org/10.1515/jnum-2012-0013

168. Hesthaven, J.S., Rozza, G., Stamm, B.: Certified Reduced Basis Methods for Parametrized Partial Differential Equations. SpringerBriefs in Mathematics. Springer, Cham; BCAM Basque Center for Applied Mathematics, Bilbao (2016)

169. Hild, P., Renard, Y.: A stabilized Lagrange multiplier method for the finite element approximation of contact problems in elastostatics. Numer. Math. **115**(1), 101–129 (2010). https://doi.org/10.1007/s00211-009-0273-z

170. Hu, Q., Chouly, F., Hu, P., Cheng, G., Bordas, S.P.A.: Skew-symmetric Nitsche's formulation in isogeometric analysis: Dirichlet and symmetry conditions, patch coupling and frictionless contact. Comput. Methods Appl. Mech. Eng. **341**, 188–220 (2018). https://doi.org/10.1016/j.cma.2018.05.024

171. Innerberger, M., Praetorius, D.: MooAFEM: an object oriented Matlab code for higher-order adaptive FEM for (nonlinear) elliptic PDEs. Appl. Math. Comput. **442**, 127731 (2023). https://doi.org/10.1016/j.amc.2022.127731

172. Jerison, D., Kenig, C.E.: The inhomogeneous Dirichlet problem in Lipschitz domains. J. Funct. Anal. **130**(1), 161–219 (1995). https://doi.org/10.1006/jfan.1995.1067

173. Jerison, D.S., Kenig, C.E.: The Dirichlet problem in non-smooth domains. Ann. Math. (2) **113**, 367–382 (1981). https://doi.org/10.2307/2006988

174. Jerison, D.S., Kenig, C.E.: The Neumann problem on Lipschitz domains. Bull. Am. Math. Soc. New Ser. **4**, 203–207 (1981). https://doi.org/10.1090/S0273-0979-1981-14884-9

175. Jerison, D.S., Kenig, C.E.: Boundary value problems on Lipschitz domains. Studies in partial differential equations, MAA Stud. Math. **23**, 1–68 (1982)

176. Johnson, C.: Numerical Solution of Partial Differential Equations by the Finite Element Method. Cambridge University Press, Cambridge (1987)

177. Juntunen, M.: On the connection between the stabilized Lagrange multiplier and Nitsche's methods. Numer. Math. **131**(3), 453–471 (2015). https://doi.org/10.1007/s00211-015-0701-1

178. Juntunen, M., Stenberg, R.: Nitsche's method for general boundary conditions. Math. Comput. **78**(267), 1353–1374 (2009). https://doi.org/10.1090/S0025-5718-08-02183-2

179. Kantorovich, L.V.: Functional analysis and applied mathematics. Usp. Mat. Nauk **3**(6(28)), 89–185 (1948)

180. Kerfriden, P., Gosselet, P., Adhikari, S., Bordas, S.P.A.: Bridging proper orthogonal decomposition methods and augmented Newton-Krylov algorithms: an adaptive model order reduction for highly nonlinear mechanical problems. Comput. Methods Appl. Mech. Eng. **200**(5–8), 850–866 (2011)

181. Kikuchi, N., Oden, J.T.: Contact Problems in Elasticity: A Study of Variational Inequalities and Finite Element Methods. SIAM Studies in Applied Mathematics, vol. 8. Society for Industrial and Applied Mathematics (SIAM), Philadelphia (1988)

182. Kikuchi, N., Song, Y.J.: Penalty-finite-element approximation of a class of unilateral problems in linear elasticity. Q. Appl. Math. **39**, 1–22 (1981)

183. Kinderlehrer, D., Stampacchia, G.: An Introduction to Variational Inequalities and Their Applications. Pure and Applied Mathematics, vol. 88. Academic Press, New York-London (1980)

184. Laborde, P., Pommier, J., Renard, Y., Salaün, M.: High-order extended finite element method for cracked domains. Int. J. Numer. Methods Eng. **64**(3), 354–381 (2005). https://doi.org/10.1002/nme.1370

185. Ladeveze, P., Leguillon, D.: Error estimate procedure in the finite element method and applications. SIAM J. Numer. Anal. **20**, 485–509 (1983)

186. Le Dret, H.: Nonlinear Elliptic Partial Differential Equations. An Introduction. Universitext. Springer, Cham (2018). https://doi.org/10.1007/978-3-319-78390-1

187. Lemaire, S.: Bridging the hybrid high-order and virtual element methods. IMA J. Numer. Anal. **41**(1), 549–593 (2021). https://doi.org/10.1093/imanum/drz056

188. Lesaint, P., Raviart, P.A.: On a finite element method for solving the neutron transport equation. In: Mathematical aspects of finite elements in partial differential equations (Proceedings of Symposia in Pure Mathematics Center, University of Wisconsin, Madison, Wisconsin, 1974), pp. 89–123. Academic Press, New York-London (1974)

189. Lions, J.L.: Quelques méthodes de résolution des problèmes aux limites non linéaires. Études mathématiques. Paris: Dunod; Paris: Gauthier-Villars. xx, 554 p. (1969)

190. Lions, J.L., Stampacchia, G.: Variational inequalities. Comm. Pure Appl. Math. **XX**, 493–519 (1967)

191. Liu, G.R., Dai, K.Y., Nguyen, T.T.: A smoothed finite element method for mechanics problems. Comput. Mech. **39**(6), 859–877 (2007)

192. Liu, M., Cai, Z.: Adaptive two-layer ReLU neural network: II. Ritz approximation to elliptic PDEs. Comput. Math. Appl. **113**, 103–116 (2022)

193. Logg, A., Wells, G.N.: DOLFIN: automated finite element computing. ACM Trans. Math. Softw. **37**(2), 20 (2010). https://doi.org/10.1145/1731022.1731030

194. Lozinski, A.: A primal discontinuous Galerkin method with static condensation on very general meshes. Numer. Math. **143**(3), 583–604 (2019). https://doi.org/10.1007/s00211-019-01067-1

195. Lützen, J.: The Prehistory of the Theory of Distributions. Studies in the History of Mathematics and Physical Sciences, vol. 7. Springer, New York-Berlin (1982)

196. Mayer, U.M., Popp, A., Gerstenberger, A., Wall, W.A.: 3D fluid-structure-contact interaction based on a combined XFEM FSI and dual mortar contact approach. Comput. Mech. **46**(1), 53–67 (2010). https://doi.org/10.1007/s00466-010-0486-0

197. Maz'ja, V.G.: Sobolev spaces. Springer Series in Soviet Mathematics. Springer, Berlin (1985). https://doi.org/10.1007/978-3-662-09922-3. Translated from the Russian by T. O. Shaposhnikova

198. McLean, W.: Strongly Elliptic Systems and Boundary Integral Equations. Cambridge University Press, Cambridge (2000)

199. Moës, N., Béchet, E., Tourbier, M.: Imposing Dirichlet boundary conditions in the extended finite element method. Int. J. Numer. Methods Eng. **67**(12), 1641–1669 (2006). https://doi.org/10.1002/nme.1675

200. Monasse, P., Perrier, V.: Orthonormal wavelet bases adapted for partial differential equations with boundary conditions. SIAM J. Math. Anal. **29**(4), 1040–1065 (1998)

201. Naumann, J.: Notes on the pre-history of Sobolev spaces. Bol. Soc. Port. Mat. **63**, 13–55 (2010)

202. Neilan, M., Salgado, A.J., Zhang, W.: Numerical analysis of strongly nonlinear PDEs. Acta Numer. **26**, 137–303 (2017). https://doi.org/10.1017/S0962492917000071

203. Nečas, J.: Les méthodes directes en théorie des équations elliptiques. Masson et Cie, Éditeurs, Paris; Academia, Éditeurs, Prague (1967)

204. Nguyen, N.C.: Model reduction techniques for parametrized nonlinear partial differential equations. In: Chouly, F., Bordas, S.P.A., Becker, R., Omnes, P. (eds.) Error Control, Adaptive Discretizations, and Applications, Part 1. Advances in Applied Mechanics (AAMS), vol. 58. Elsevier, Amsterdam (2024). https://doi.org/10.1016/bs.aams.2024.03.005

205. Nguyen, V.P., Anitescu, C., Bordas, S.P., Rabczuk, T.: Isogeometric analysis: an overview and computer implementation aspects. Math. Comput. Simul. **117**, 89–116 (2015)

206. Nguyen, V.P., Rabczuk, T., Bordas, S.P.A., Duflot, M.: Meshless methods: a review and computer implementation aspects. Math. Comput. Simul. **79**(3), 763–813 (2008)

207. Nguyen-Xuan, H., Rabczuk, T., Bordas, S., Debongnie, J.F.: A smoothed finite element method for plate analysis. Comput. Methods Appl. Mech. Eng. **197**(13–16), 1184–1203 (2008)

208. Nicaise, S., Semin, A.: Density and trace results in generalized fractal networks. ESAIM, Math. Model. Numer. Anal. **52**(3), 1023–1049 (2018). https://doi.org/10.1051/m2an/2018021

209. Nitsche, J.: Über ein Variationsprinzip zur Lösung von Dirichlet-Problemen bei Verwendung von Teilräumen, die keinen Randbedingungen unterworfen sind. Abh. Math. Sem. Univ. Hamburg **36**, 9–15 (1971)

210. Nochetto, R.H., Siebert, K.G., Veeser, A.: Theory of adaptive finite element methods: An introduction. In: Multiscale, nonlinear and adaptive approximation. Dedicated to Wolfgang Dahmen on the Occasion of His 60th Birthday, pp. 409–542. Springer, Berlin (2009). https://doi.org/10.1007/978-3-642-03413-8_12

211. Oden, J.T., Kim, S.: Interior penalty methods for finite element approximations of the Signorini problem in elastostatics. Comput. Maths. Appl. **8**(1), 35–56 (1982)

212. Oden, J.T.: Historical comments on finite elements. In: A history of Scientific Computing (Princeton, NJ, 1987), ACM Press History Series, pp. 152–166. ACM, New York (1990)

213. Oden, J.T., Prudhomme, S.: Estimation of modeling error in computational mechanics. J. Comput. Phys. **182**(2), 496–515 (2002). https://doi.org/10.1006/jcph.2002.7183

214. Papež, J.: Algebraic error in numerical PDEs and its estimation. In: Chouly, F., Bordas, S.P.A., Becker, R., Omnes, P. (eds.) Error Control, Adaptive Discretizations, and Applications, Part 1. Advances in Applied Mechanics (AAMS), vol. 58. Elsevier, Amsterdam (2024). https://doi.org/10.1016/bs.aams.2024.04.002

215. Papež, J., Vohralík, M.: Inexpensive guaranteed and efficient upper bounds on the algebraic error in finite element discretizations. Numer. Algorithms **89**(1), 371–407 (2022). https://doi.org/10.1007/s11075-021-01118-5

216. Persson, P.O., Strang, G.: A simple mesh generator in MATLAB. SIAM Rev. **46**(2), 329–345 (2004). https://doi.org/10.1137/S0036144503429121

217. Peskin, C.S.: The immersed boundary method. Acta Numer. **11**, 479–517 (2002). https://doi.org/10.1017/S0962492902000077

218. Picasso, M.: Numerical study of the effectivity index for an anisotropic error indicator based on Zienkiewicz–Zhu error estimator. Commun. Numer. Methods Eng. **19**(1), 13–23 (2003). https://doi.org/10.1002/cnm.546

219. Quarteroni, A.: Numerical Models for Differential Problems. MS&A, Modeling and Simulation: An Application, vol. 16, 3rd edn. Springer, Cham (2018). https://doi.org/10.1007/978-3-319-49316-9

220. Quarteroni, A., Valli, A.: Numerical Approximation of Partial Differential Equations. Springer Series in Computational Mathematics, vol. 23. Springer, Berlin (1994)

221. Quarteroni, A., Valli, A.: Domain Decomposition Methods for Partial Differential Equations. Numerical Mathematics and Scientific Computation. The Clarendon Press, Oxford University Press, New York (1999)

222. Raviart, P.A., Thomas, J.M.: Introduction à l'analyse numérique des équations aux dérivées partielles. Collection Mathématiques Appliquées pour la Maîtrise. [Collection of Applied Mathematics for the Master's Degree]. Masson, Paris (1983)

223. Renard, Y.: Generalized Newton's methods for the approximation and resolution of frictional contact problems in elasticity. Comput. Methods Appl. Mech. Eng. **256**, 38–55 (2013). https://doi.org/10.1016/j.cma.2012.12.008

224. Renard, Y., Poulios, K.: GetFEM: automated FE modeling of multiphysics problems based on a generic weak form language. ACM Trans. Math. Softw. **47**(1), Art. 4, 31 (2021). https://doi.org/10.1145/3412849

225. Repin, S.: A Posteriori Estimates for Partial Differential Equations. Radon Series on Computational and Applied Mathematics, vol. 4. de Gruyter, Berlin (2008). https://doi.org/10.1515/9783110203042

226. Repin, S., Sauter, S., Smolianski, A.: A posteriori error estimation for the Dirichlet problem with account of the error in the approximation of boundary conditions. Computing **70**(3), 205–233 (2003). https://doi.org/10.1007/s00607-003-0013-7

227. Repin, S.I.: A posteriori error identities and estimates of modelling errors. In: Chouly, F., Bordas, S.P.A., Becker, R., Omnes, P. (eds.) Error Control, Adaptive Discretizations, and Applications, Part 1. Advances in Applied Mechanics (AAMS), vol. 58. Elsevier, Amsterdam (2024). https://doi.org/10.1016/bs.aams.2024.03.006

228. Rognes, M.E., Logg, A.: Automated goal-oriented error control. I: stationary variational problems. SIAM J. Sci. Comput. **35**(3), c173–c193 (2013). https://doi.org/10.1137/10081962X

229. Rüter, M., Gerasimov, Y., Stein, E.: Goal-oriented explicit residual-type error estimates in XFEM. Comput. Mech. **52**(2), 361–376 (2013)

230. dos Santos, N.D., Gerbeau, J.F., Bourgat, J.F.: A partitioned fluid-structure algorithm for elastic thin valves with contact. Comput. Methods Appl. Mech. Eng. **197**(19–20), 1750–1761 (2008). https://doi.org/10.1016/j.cma.2007.03.019

231. Savaré, G.: Regularity results for elliptic equations in Lipschitz domains. J. Funct. Anal. **152**(1), 176–201 (1998). https://doi.org/10.1006/jfan.1997.3158

232. Sayas, F.J., Brown, T.S., Hassell, M.E.: Variational Techniques for Elliptic Partial Differential Equations. CRC Press, Boca Raton (2019). https://doi.org/10.1201/9780429507069

233. Schwartz, L.: Théorie des distributions. Publications de l'Institut de Mathématique de l'Université de Strasbourg, No. IX-X. Hermann, Paris (1966)

234. Scott, L.R., Zhang, S.: Finite element interpolation of nonsmooth functions satisfying boundary conditions. Math. Comput. **54**(190), 483–493 (1990). https://doi.org/10.2307/2008497
235. Sofonea, M.: Well-Posed Nonlinear Problems. A Study of Mathematical Models of Contact. Advances in Applied Mathematics, vol. 50. Birkhäuser, Cham (2023). https://doi.org/10.1007/978-3-031-41416-9
236. Steinbach, O.: Numerical Approximation Methods for Elliptic Boundary Value Problems. Springer, New York (2008). https://doi.org/10.1007/978-0-387-68805-3
237. Stenberg, R.: On some techniques for approximating boundary conditions in the finite element method. J. Comput. Appl. Math. **63**(1–3), 139–148 (1995). https://doi.org/10.1016/0377-0427(95)00057-7
238. Strang, G., Fix, G.J.: An Analysis of the Finite Element Method. Prentice-Hall Series in Automatic Computation. Prentice-Hall, Englewood Cliffs (1973)
239. Sukumar, N., Tabarraei, A.: Conforming polygonal finite elements. Int. J. Numer. Methods Eng. **61**(12), 2045–2066 (2004)
240. Szabó, B., Babuška, I.: Finite Element Analysis. Formulation, Verification and Validation, 2nd updated and revised edition edn. Wiley Series in Computational Mechanics. John Wiley & Sons, Hoboken (2021). https://doi.org/10.1002/9781119426479
241. Tartar, L.: An Introduction to Sobolev Spaces and Interpolation Spaces. Lecture Notes of the Unione Matematica Italiana, vol. 3. Springer, Berlin; UMI, Bologna (2007)
242. Thomée, V.: Galerkin Finite Element Methods for Parabolic Problems. Springer Series in Computational Mathematics, vol. 25. Springer, Berlin (1997)
243. Urquiza, J.M., Garon, A., Farinas, M.I.: Weak imposition of the slip boundary condition on curved boundaries for Stokes flow. J. Comput. Phys. **256**, 748–767 (2014). https://doi.org/10.1016/j.jcp.2013.08.045
244. Verfürth, R.: A Posteriori Error Estimation Techniques for Finite Element Methods. Numerical Mathematics and Scientific Computation. Oxford University Press, Oxford (2013)
245. Warburton, T., Hesthaven, J.S.: On the constants in hp-finite element trace inverse inequalities. Comput. Methods Appl. Mech. Eng. **192**(25), 2765–2773 (2003). https://doi.org/10.1016/S0045-7825(03)00294-9
246. Winter, M., Schott, B., Massing, A., Wall, W.A.: A Nitsche cut finite element method for the Oseen problem with general Navier boundary conditions. Comput. Methods Appl. Mech. Eng. **330**, 220–252 (2018). https://doi.org/10.1016/j.cma.2017.10.023
247. Wohlmuth, B.I.: A mortar finite element method using dual spaces for the Lagrange multiplier. SIAM J. Numer. Anal. **38**(3), 989–1012 (2000). https://doi.org/10.1137/S0036142999350929
248. Wohlmuth, B.I.: Variationally consistent discretization schemes and numerical algorithms for contact problems. Acta Numer. **20**, 569–734 (2011)
249. Zienkiewicz, O.C., Taylor, R.L.: The Finite Element Method. vol. 1, 5th edn. Butterworth-Heinemann, Oxford (2000)
250. Zienkiewicz, O.C., Zhu, J.Z.: A simple error estimator and adaptive procedure for practical engineering analysis. Int. J. Numer. Methods Eng. **24**(2), 337–357 (1987)

Index

A

Abstract a priori error bound, 68, 92
Abstract error estimate, 106
Adaptive finite element method, 118
Adaptive mesh refinement, 118
Anisotropic mesh, 123
A posteriori error estimator, 117
Approximation, 51
Approximation error, 53
A priori error estimate, 68, 69, 94
Aubin-Nitsche duality trick, 76

B

Becker and Rannacher technique, 120
Becker, Estecahandy, and Trujillo technique, 121
Best approximation error, 53
Bi-Lipschitz homeomorphism, 16
Boundary value problem, 12, 102, 113

C

Cholesky, 72, 95
Closed-form solution, 22, 103
Column vector, 71
Commuting property, 59
Compatibility condition, 33, 35, 41, 74
Conformity (H^1-conformity), 58
Conjugate gradient, 72, 74, 95
Consistency, 87
Continuous functions, 5
Convex domain, 17
Convex hull, 54
Convex polygon, 44
Convex quadratic functional, 31
Coulomb boundary condition, 47

Crack, 17
Cusp, 17

D

Damaged mesh, 123
Degenerate simplex, 54, 124
Degrees of freedom, 71
Deny-Lions Theorem, 36
Diffusion equation, 12
Dirac delta function, 8
Direct nodal imposition, 64
Dirichlet boundary condition, 13, 21, 47, 64, 82, 113
Discontinuous Galerkin Interior Penalty, 85
Discrete cone, 105
Discrete error, 53, 116
Discrete lifting, 64, 73
Discrete trace inequality, 86
Distribution, 6
Divergence, 5
Divergence Theorem, 19
Domain decomposition, 47
Dual singular function, 44
Dual-weighted Residuals, 120
Dual space, 20

E

Edge, 55, 56
Efficiency, 119
Elliptic partial differential equation, 46, 75
Energy functional, 83
Energy minimization, 13, 68
Equilibrated fluxes estimator, 124
Error estimate, 59, 64
Essential boundary condition, 13

Existence, 14
Extraction formula, 44

F
Face, 55, 56
Facet, 82
Finite element method, 63, 79, 111
First order optimality condition, 83
Fluid-structure interaction, 47
Function, 4

G
Galerkin method, 51
Galerkin procedure, 63
Gauss Theorem, 19
Goal-oriented error estimation, 120
Goal-oriented error estimator, 124
Gradient, 5
Gradient recovery estimator, 124
Green formula, 19, 82
Green Theorem, 40

H
Hadamard, 14
Heaviside function, 8
Hierarchical error estimator, 124
Hilbert space, 14, 20, 88
Hybrid high order method, 103

I
Inclusion, 76
Incomplete Nitsche method, 83
Inner product, 20, 23, 43
Interface condition, 47
Interface problem, 48

J
Jump, 117

K
Krylov, 95

L
Lagrange finite element, 52
Lagrange finite element space, 57, 66, 83, 114
Lagrange interpolant, 64

Lagrange interpolator, 58
Laplace equation, 12
Laplace operator, 19
Lax-Milgram Theorem, 88
Lebesgue space, 23
Lifting, 14, 26, 27
Linear system, 63, 65
Lipschitz domain, 16, 17, 46
Lipschitz function, 18
Locally integrable funcions ($L^1_{loc}(\Omega)$), 7
Lower bound, 119
L-shape domain, 44
Lu decomposition, 95

M
Mathematical model, 11
Matrix, 63, 65
Matrix formulation, 72
Mesh, 52, 60
Mesh generation, 123
Meshing, 60
Mesh regular in Ciarlet's sense, 57
Mesh size, 52, 57
Minimization problem, 31, 41
Mixed boundary conditions, 47
Mixed method, 75
Modelling error, 123
Multi-index, 4

N
Neumann boundary condition, 15, 32, 47, 73
Nitsche, 109
Nitsche matrix, 95
Nitsche method, 75, 85, 109
Nitsche parameter, 82
Nodal basis, 52
Node, 54, 56
Nonconvex polygon, 44
Norm, 23
Normal, 18
Numerical approximation method, 60
Numerical error, 63, 73, 123
Numerical simulation, 11
Numerical test, 74

O
Open Mapping Theorem, 27
Optimal error bound, 70
Optimal error estimate, 74

P

Partial derivative, 4
Partial differential equation, 46
Partitioning procedure, 72
Penalty-free Nitsche method, 85
Penalty method, 75
Piecewise polynomial approximation, 52
Pipeline, 12, 51, 63, 79, 99, 111
Poincaré-Friedrichs inequality, 14, 29
Poincaré inequality, 37
Poisson's problem, 12, 21, 32, 113
Postprocessing, 111

R

Rademacher's Theorem, 18
Reentrant angle, 44
Reentrant corner, 44
Refined mesh, 122
Regular distribution, 7
Regularity, 15, 42, 44, 46
Regularity estimate, 45
Regular part, 44
Reliability, 119
Residual, 116
Residual error estimator, 117
Riesz-Fréchet Representation Theorem, 14, 20
Ritz-Galerkin method, 51
Robin-Fourier boundary condition, 47
Roundoff, 63

S

Scikit-fem, 95
Segment, 55
Shape function, 58
Shape-regularity, 57
Signorini boundary condition, 47, 108
Signorini problem, 102
Simplex, 52, 54, 56
Simplicial mesh, 52, 54, 55, 114
Singular function, 44
Singularity, 15, 44, 46
Skew-symmetric Nitsche method, 85

Sobolev embedding Theorem, 43
Sobolev norm, 43
Sobolev semi-norm, 59
Sobolev space, 24, 43, 46
Space ($H(\mathrm{div}; \Omega)$ space), 40
Stability, 14
Stability bound, 29
Stampacchia's Theorem, 104
Stress intensity factors, 45
Strong-weak equivalence, 31, 39
Support, 5
Symmetric Nitsche method, 83

T

Test functions, 5
Tetrahedron, 55
Trace mapping, 24
Trace space, 25
Trace Theorem, 24
Tresca boundary condition, 47
Triangle, 55

U

Uniqueness, 14
Upper bound, 119

V

Variational approximation, 51
Variational equation, 23
Variational inequality, 104, 105
Variational lemma, 7
Vector-valued distribution, 7
Vector-valued function, 5
Vertex, 54, 55

W

Weak form, 23, 26, 35
Weak formulation, 13, 103
Well-posedness, 14, 29, 38, 64, 67, 88